PROBING THE OCEAN FOR SUBMARINES

A HISTORY OF THE AN/SQS-26 LONG-RANGE, ECHO-RANGING SONAR

A Memoir by

Thaddeus G. Bell

**Second Edition
2010**

Approved for public release; distribution is unlimited.

PROBING THE OCEAN FOR SUBMARINES

Second Edition

This book is dedicated to William A. Downes, who initiated the AN/SQS-26 program in 1955 and who directed with wisdom and perseverance the ensuing developments until his retirement from the Naval Underwater Systems Center in 1971.

FOREWORD

The first edition of Thad Bell's memoir was originally published by the Naval Undersea Warfare Center Division, Newport, RI, in 2003. With the publication of this revised edition, the significant historical material in the original book is available for general use by academia, research laboratories, and fleet units interested in antisubmarine warfare and the principles of sonar.

This is the story of one of the most challenging programs of the Cold War era. Combining the knowledge and craftsmanship of engineering, naval architecture, ocean science, and operational expertise, the AN/SQS-26 program's success was a key factor in the U.S. Navy's quest for ASW superiority. As with any undertaking of this scale, there needed to be a "hero," an individual within the organization who had the vision, in-depth knowledge, perseverance, and voice to steer the sonar program through the difficult design, development, testing, and operational employment stages. That hero was Thaddeus G. Bell at the Naval Underwater Systems Center, New London, CT.

Richard F. Pittenger
Rear Admiral, U.S. Navy (Retired)
November 2010

PREFACE

Since my retirement in 1985 from the Naval Underwater Systems Center in New London, Connecticut (now merged with the Naval Undersea Warfare Center (NUWC) in Newport, Rhode Island), a number of colleagues and friends have suggested that I write a history of my activities in the AN/SQS-26 development program. The press of requests to contribute to ongoing sonar projects, however, made it difficult to find the time. With the end of the Cold War in 1991, my efforts on those projects began to taper off, and by the mid-nineties, I seriously considered beginning the history. I felt a sense of urgency to undertake it while many of the early participants were still available.

I decided early on that I would not attempt a *classified* history. The difficulty that a potential reader would encounter when attempting to acquire access to a classified publication would largely defeat the purpose of making this information conveniently available to a wide audience. Even general knowledge of the *existence* of a classified document tends to fade rather rapidly with time. Although some of the source material could obviously be declassified within the existing guidelines, it appeared feasible to use unclassified excerpts from classified documents for the remainder. These excerpts would still provide a large amount of interesting and valuable historical information.

On 17 April 1995, Executive Order 12958 was issued regarding the declassification of all documents more than 25 years old, unless justification for exceptions was provided. I thought this would simplify my task since the major development work on the SQS-26 sonar occurred prior to 1975.[*] Thus, it seemed that key SQS-26 documents would meet the 25-year age requirement for declassification, yet would not meet the basic criteria that (in my judgment) would establish an exception to being declassified. With this reasoning, I believed that I would be able to obtain access to most SQS-26 documents without setting up a special project to establish a "need to know."

---

[*]In referring to the "SQS" series of sonars from this point forward, I shall follow the usual convention of dropping the "AN-" prefix that accompanies the formal names.

Then, on 8 April 1996, the Navy Department issued OPNAV Instruction 5513.16A, containing guidelines for *exceptions* to the declassification order. The exceptions included a great deal of source information on the SQS-26 that was contained in documents that I would need to consult.

After I began my writing, however, I received some unexpected news regarding declassification actions. In May 1999, a representative of the Naval Sea Systems Command (NAVSEA 09) visited NUWC Division Newport for a review of classified documents more than 25 years old. During the process, he declassified 30 documents on the SQS-26. Although I was disappointed that the number was not greater, many of those documents contained important historical information on system performance that I could not have otherwise presented here.

It was still necessary to consult a considerable number of classified documents for relevant unclassified information. The parts of the documents that were clearly unclassified were not difficult to identify, and it was also not unusual to find that many of the pages in such documents were stamped "unclassified" by the originators. Moreover, there was much in these documents of a purely administrative nature that was of historical value. To obtain access to these publications, however, required a clearance and an established "need to know," which meant that a formally funded project, preferably with NUWC Division Newport, would have to be established.

During informal discussions about this possibility with a number of personnel at the Division, James Donald of the Special Projects Program Office (who was one of the early participants in SQS-26 development) believed that he could arrange to acquire funding for the written history project. Because I already had office space and ongoing work at Analysis and Technology (A&T) — now Anteon Corporation — in North Stonington, Connecticut, it was decided that A&T would provide the needed administrative support and classified material storage. Later, when I discussed the project with the Executive Director of NUWC Division Newport, Juergen Keil, who had played a key role in the later stages of the SQS-26 development work, I found him to be very supportive.

Beginning the history did not occur until after the New London Laboratory was closed on 4 October 1996 and moved to NUWC Division Newport. Because material in the New London Library was considerably thinned out as a result of this move, I feared that few of the source documents would still be available. However, upon acquiring access to the Newport library in June 1997, I was immensely relieved to find an enormous collection of SQS-26 technical memoranda and formal reports listed in the computer file of holdings. My first request for a readout resulted in some 1800 available SQS-26 documents. The librarians at New London during the SQS-26 development program had done a remarkable job of identifying each document that should be included in the SQS-26 file, even when the term "SQS-26" did not appear in the publication. Thus, even relevant documents generated prior to the assignment of the SQS-26 name to the project found their way into the collection.*

Limiting the size of the final document to no more than a few hundred pages of history meant that I would have to reduce the 1800 documents that I had found down to a group of core documents. The core documents would be those that (in my judgment) would be suitable references for the history, considering the constraints on the final document size and the time remaining to complete the project. With the cooperation of Mary Barravecchia (the Newport head librarian), Charles Logan was assigned to help me gather the information that I needed. He initially provided me with computer listings of SQS-26 documents by year, afterward retrieving only those that I wished to inspect more closely.

In the end, 325 key documents constituted the major source material. Since I would need continuing access to these 325 documents, the copies were shipped to A&T in North Stonington. Other material that supplemented these documents included (1) relevant NUWC publications generated outside the SQS-26 program, (2) selected documents obtained from the Naval Research Laboratory, (3) publications in the open literature, (4) informal papers and notes that I had accumulated over the years, (5) conversations with surviving participants, and (6) my own memory of events. What I present will be in the

---

*This result must be attributed, in great measure, to the conscientiousness of key New London librarians Dorothy Morris and Ruth Maples.

style of a memoir and therefore will emphasize the part of the development story with which I was most closely involved. As a consequence, there will be some inevitable gaps in the account. On the other hand, my involvement in the SQS-26 development was such that the reader will be presented with firsthand knowledge of many significant milestones.

ACKNOWLEDGMENTS

I am greatly indebted to the following former colleagues who reviewed the manuscript and provided corrections, comments, and valuable supplementary information: Russell E. Baline; Dr. Bradford A. Becken; Dr. David G. Browning; Dr. Robert P. Chapman; Dr. Aldo "Gene" DiLoretto; Harold J. Doebler; John J. Hanrahan; Walter C. Hay; Juergen G. Keil; Kyrill V. Korolenko; Dr. Robert H. Mellen; John Merrill; Rear Admiral William A. Myers, USN (Retired); Michael Pastore; Rear Admiral Richard F. Pittenger, USN (Retired); Dr. Charles H. Sherman; John R. Snow; Dr. James L. Stewart; and Harry J. Tucker. Special appreciation is extended to former NUWC Division Newport Executive Director Juergen Keil, who approved the funding that permitted publication of this book.

In addition, James B. Donald and Cynthia B. Straney provided essential liaison with NUWC Division Newport. The NUWC head librarian Mary N. Barravecchia made arrangements for access to SQS-26 technical documents, and librarian Charles R. Logan provided much valuable working-level assistance in document retrieval. Dr. Burton G. Hurdle of the Naval Research Laboratory was most cooperative in having relevant Naval Research Laboratory documents declassified and then forwarded to me. I am also indebted to Karen Holt of NUWC and Linda Turner of Systems Resource Management Inc. for their editing of the manuscript and to Dana Gardner of Systems Resource Management Inc. for the cover design.

TABLE OF CONTENTS

Chapter		Page
	LIST OF ILLUSTRATIONS	viii
	LIST OF TABLES	ix
	LIST OF ABBREVIATIONS AND ACRONYMS	xi
1	INTRODUCTION	1
	Definition of Long Range	1
	Scope of SQS-26 History	2
	System Longevity	3
2	HISTORICAL BACKGROUND	5
	Echo-Ranging Sonar Before 1955	5
	Post-War Perception of the Soviet Submarine Threat	10
	Post-War Research on Long-Range Sound Paths	14
	Fruition of Research Programs	16
3	LAUNCHING A LONG-RANGE ACTIVE SONAR PROGRAM: EARLY CONCEPT FORMULATION	19
	Concept Formulation Studies at the Navy Underwater Sound Laboratory	19
	Interactions with the Outside World	25
	BRASS Experiments	36
	BRASS II	38
	Normal-Incidence Bottom Loss Survey Concept	41
	Demise of Scout Ship Concept	42
	Implementation Details for the Conceptual Design	43
	Procurement Plans	48
	Importance of Systems Engineering Function to SQS-26	52
	Analysis of SQS-26 Systems for Open-Ocean Search	55
	Development Status at the End of 1960	58
4	FULL-SCALE EXPERIMENTATION AND DEVELOPMENT	61
	Two SQS-26 Experimental Systems	61
	Management Concern with Equipment Reliability and Performance	68
	NUSL Concern About Maintaining and Operating Production Systems	70
	Further XN-1 and XN-2 Testing and Analysis	71
	Operational Evaluation of the XN-2	73
	Navy Reaction to the Operational Evaluation Failure	75
	Expansion of the SQS-26 Program	78

TABLE OF CONTENTS (Cont'd)

Chapter		Page
5	PROTOTYPE TESTING	91
	Testing the SQS-26 (BX) Production System	91
	Testing the XN-2 Major Retrofit Prototype	91
	Unrecognized Failures in the Production Equipment: The SEA Teams	94
	Testing the SQS-26 (CX) Production System	96
6	SUPPORTING RESEARCH AND DEVELOPMENT	99
	Tracor Contract	99
	Biological Reverberation	101
	Array Receiving Phases	102
	Display Resolution	102
	Marine Geophysical Survey Program	103
	Attenuation Coefficient	107
	Shipboard Prediction Methodology	107
	SQS-26 Display Testing at NUSL	109
	Scattering Strengths in SQS-26 Test Area "B"	109
	Joint Oceanographic Acoustic and System Tests	110
7	THE RUBBER DOME WINDOW	111
	Steel Dome Problems	111
	Rubber Dome Window Proposal	111
	Echo-Ranging Performance with Rubber and Steel Dome Windows	114
	Expected Operational Impact of the Rubber Dome Window	116
8	EQUIPMENT OPERATION AND TACTICAL EMPLOYMENT	121
	Early Concern About Operator Training	121
	Expanding the SQS-26 Operating Doctrine Program at NUSL	124
9	FLEET PERFORMANCE	129
	Importance of Information Related to Fleet Performance	129
	September 1965: Observations of SQS-26 Performance in a Fleet Exercise	130
	July 1966: Free-Play Success with the Convergence Zone Mode	131
	May 1968: Feasibility Studies of Convergence Zone Applications in the Mediterranean Sea	132
	November 1968: Convergence Zone Sweep of the Tyrrhenian Sea with Two Ships	139

TABLE OF CONTENTS (Cont'd)

Chapter		Page
9	March 1969: Carrier Screening Exercise Off San Diego	143
	July 1969: Successful Sweep Operation in the Convergence Zone	145
	July 1969: Convergence Zone Exercise Vectoring a Pouncer into Attack Range in the Tyrrhenian Sea	146
	August 1970: Bottom Bounce Performance Tests in the Mediterranean Sea	146
	April 1971: Free-Play Convergence Zone Experience on *Horne* for HOLDEX 2-71 in the Pacific Ocean	147
	August 1971: Shallow-Water Testing on the Tunisian Shelf	148
	September 1971: Validation of the NUSC Acoustic Province Chart	149
	September 1971: Semifree-Play Convergence Zone Exercise in the Mediterranean Sea	150
	October 1971: Free-Play Convergence Zone Detection and Attack Operations with Two Ships	150
	December 1971: Convergence Zone Contact During a Random Encounter with a U.S. Submarine	151
	January 1972: Continued Success by *Belknap*	151
	April 1972: *Sims* Convergence Zone Performance with a Soviet Submarine as the Target	152
	April 1972: *Sims* Random Encounters with U.S. Nuclear Submarines	153
	July 1972: Decline in Reported Convergence Zone Contacts	154
	May 1973: Coordinated Operations in the Pacific Ocean	154
	August 1973: SHAREM XVI (MD) Convergence Zone Results in the Mediterranean Sea	156
	December 1973: Analysis of Atlantic Fleet's Integrated Escort Tactical Development Program	157
	December 1974: Observations of Incorrect System Operation	163
	October 1975: LAMPS III Testing in the North Atlantic	165
	April 1976: *Connole* and the ASW Squadron in the Mediterranean Sea	165
	Final Report on Fleet Results with the SQS-26	167
	Significance of Fleet Performance Observations	168
10	CONCLUSIONS	171
	Accomplishments of the SQS-26 Program	171
	Contributions to Program Success	172

TABLE OF CONTENTS (Cont'd)

Chapter   **Page**

10	International Situation	174
	Navy Laboratory System	175
	Upper Echelon Management Support	179
	Policy on Contracting with Private Industry	179
	Summary of Contributions to Program Success	180
	Trends in ASW Beyond 1975	180
11	ANNOTATED ENDNOTES	185
	APPENDIX A — CHRONOLOGY OF EVENTS INFLUENCING THE DEVELOPMENT AND APPLICATION OF THE SQS-26 SONAR	A-1
	APPENDIX B — LIST OF PERSONNEL HAVING AN IMPACT ON THE SQS-26 PROGRAM	B-1
	INDEX	I-1

LIST OF ILLUSTRATIONS

Figure   **Page**

1	Deep-Ocean Sound Paths Exploitable by the SQS-26 Long-Range, Echo-Ranging Sonar	2
2	Display on Scanning Sonar Developed by HUSL During World War II	8
3	The Shadow Zone Below a 200-Foot Isothermal Layer	12
4	Three Early Pioneers of Active Sonar	17
5	Echo Excess Versus Frequency at Two Target Ranges	22
6	BRASS II Experimental Array	39
7	Notional Formation of Long-Range Active Sonar Ships for ASW Search of a Large Ocean Area	57
8	High-Intensity Zone Occurring in Bottom-Reflected Sound When Bottom Depth Is Not Sufficient for Convergence Zone Formation	85
9	Systematic Dependence on Time of Day in the Gulf of Mexico Due To Biological Reverberation	86
10	Problems Encountered by SEA Teams	95
11	Problems Existing upon Departure of SEA Teams	96
12	Components of the SQS-26 (CX) in the Early 1970's	98

LIST OF ILLUSTRATIONS (Cont'd)

Figure		Page
13	Three-Segment Norfolk to Gibraltar Route Selected for Computations of SQS-26 Detection Performance	117
14	Notional Escort Spacing for Four SQS-26 Ships	117
15	Direct Path and Convergence Zone Detection Coverage of Shallow-Water Submarines by the SQS-26 in the Mediterranean Sea	135
16	Ray Paths During the Transition Months from Convergence Zone to Direct Path and Back Again	136
17	Thaddeus Bell (the Author) During a Visit to USS *Little Rock* (CLG-4) in the Mediterranean Sea	142
18	Detection-to-Opportunity Ratio Versus Closest Point of Approach for Both SQS-26 and Older Sonars	160
19	Detection-to-Opportunity Ratio Versus Range Normalized to the 50% Probability Detection Range	162
20	ASW Squadron at Naples in 1976	167

LIST OF TABLES

Table		Page
1	Detection Range Statistics Computed for Environment on Norfolk to Gibraltar Route	118
2	July 1969 Sweep Operation in the Convergence Zone	145
3	Distribution of Opportunity Hours as a Function of Range Normalized to 50% Probability Detection Range	163

LIST OF ABBREVIATIONS AND ACRONYMS

AAW	Anti-air warfare
AIP	Air-independent propulsion
AMOS	Acoustic, meteorological, oceanographic survey
ASROC	Antisubmarine rocket
ASW	Antisubmarine warfare
ASWFORSIXTHFLT	ASW Force, Sixth Fleet
ASWGRUONE	ASW Group One
ATP	Allied tactical publication
AUWE	Admiralty Underwater Weapons Establishment
BB	Bottom bounce
BRASS	Bottom-Reflected Active Sonar System
BuPers	Bureau of Personnel (Navy)
BuShips	Bureau of Ships (later NAVSEA)
CAPT	Captain
CDR	Commander
CG	Guided missile cruiser
CGN	Guided missile cruiser (nuclear propulsion)
CIC	Combat information center
CINCLANT	Commander-in-Chief, Atlantic
CNM	Chief of Naval Material
CNO	Chief of Naval Operations
COMASWFORSIXTHFLT	Commander, ASW Force, Sixth Fleet
COMDESDEVGRU	Commander, Destroyer Development Group
COMNAVSURFLANT	Commander, Naval Surface Forces, Atlantic Fleet
COMOPTEVFOR	Commander, Operational Test and Evaluation Force
COMSUBLANT	Commander, Submarine Force, Atlantic Fleet
CP	Coded pulse
CPA	Closest point of approach
CRT	Cathode ray tube
CTU	Commander, task unit
CV	Aircraft carrier
CVS	Aircraft carrier (ASW support)
CVY	Charlie/Victor/Yankee
CW	Continuous wave
CZ	Convergence zone
DD	Destroyer
DE	Destroyer escort
DEG	Guided missile escort ship
DELTIC	Delay line time compression

LIST OF ABBREVIATIONS AND ACRONYMS (Cont'd)

DESRON	Destroyer squadron
DL	Destroyer leader
DLG	Guided missile frigate
DoD	Department of Defense
ESM	Electronic warfare support measure
FF	Frigate
FFG	Guided missile frigate
FM	Frequency modulated
FSK	Frequency shift keying
FY	Fiscal year
GE	General Electric
HEN	Hotel/Echo/November
HOLDEX	(Submarine contact) holding exercise
HUKASWEX	Hunter-killer ASW exercise
HUSL	Harvard Underwater Sound Laboratory
IEP	Integrated Escort Program
IRE	Institute of Radio Engineers
JOAST	Joint Oceanographic Acoustic and System Test
LAMPS	Light airborne multipurpose system
LANCORT	Atlantic (ASW) escort (exercise)
LAPS	Louis Allis power supply
LORAD	Long-range active detection
MAD	Magnetic anomaly detector
MGS	Marine Geophysical Survey
MI	Mutual interference
MILOC	Military oceanography conference
MIT	Massachusetts Institute of Technology
MRF	Major retrofit
MTBF	Mean time between failure
NATO	North Atlantic Treaty Organization
NAVMAT	Naval Material Command
NAVOCEANO	Naval Oceanographic Office
NAVSEA	Naval Sea Systems Command (formerly BuShips)
NAVSHIPS	Naval Ship Systems Command
NEL	Naval Electronics Laboratory
NRL	Naval Research Laboratory
NSIA	National Security Industrial Association
NUSC	Naval Underwater Systems Center
NUSL	Navy Underwater Sound Laboratory

LIST OF ABBREVIATIONS AND ACRONYMS (Cont'd)

NUWC	Naval Undersea Warfare Center
NWIP	Naval warfare interim publication
ODN	Own Doppler nullifier
ODT	Omnidirectional transmission
ONI	Office of Naval Intelligence
ONR	Office of Naval Research
OPEVAL	Operational evaluation
OPNAV	Office of Chief of Naval Operations
OPTEVFOR	Operational Test and Evaluation Force
OTC	On-site tactical commander
PDC	Practice depth charge
PDT	Processed directional transmission
PPI	Plan position indicator
PRN	Pseudorandom noise
PSR	Predicted sonar range
RAYMODE	Mathematical model for underwater sound propagation computations
RDW	Rubber dome window
SACLANT	Supreme Allied Command, Atlantic
SAR	Search and rescue
SEA	Sonar evaluation and assistance
SHAREM	Ship ASW Readiness/Effectiveness Measurement
SIMAS	Sonar *in-situ* mode assessment system
SOFIX	Sonar fix ("get-well" program for SQS-26 development)
SOSUS	Sound surveillance undersea system
SS	Submarine
SSBN	Fleet ballistic missile submarine, nuclear propulsion
SSGN	Submarine, guided missile, nuclear propulsion
SSI	Sector scan indicator
SSN	Submarine, fast attack, nuclear propulsion
SURTASS	Surveillance towed array sonar system
SW	Southwest
TASS	Towed Array Sonar System
TDA	Technical development agent
TMA	Target motion analysis
TTD	Technical test director
UK	United Kingdom
USAG	Underwater Sound Advisory Group

LIST OF ABBREVIATIONS AND ACRONYMS (Cont'd)

USN	United States Navy
USSR	Union of Soviet Socialist Republics
VP	ASW land-based marine patrol aircraft
VS	ASW carrier-based aircraft
WHOI	Woods Hole Oceanographic Institution
WSEG	Weapon Systems Evaluation Group

PROBING THE OCEAN FOR SUBMARINES

A HISTORY OF THE AN/SQS-26 LONG-RANGE, ECHO-RANGING SONAR

A Memoir by

Thaddeus G. Bell

Second Edition
2010

CHAPTER 1
INTRODUCTION*

DEFINITION OF LONG RANGE

The title of this book indicates that the account will constitute a "history of the AN/SQS-26 long-range, echo-ranging sonar." The reader might ask, What exactly is meant by *long range* in this context? As defined here, the term pertains to a combination of two SQS-26 capabilities that were not present in predecessor production sonars: (1) over-the-horizon detection and (2) the use of three major deep-ocean sound paths.

Over-the-Horizon Capability

The visual horizon, as perceived from a ship's bridge 50 feet above the waterline, is about 9 nautical miles, or 18 kiloyards. The SQS-23 (predecessor system to the SQS-26) had a nominal direct path detection range of about 10 kiloyards. The goal of the original SQS-26 design was a typical bottom bounce range of 20 kiloyards, which would clearly provide an over-the-horizon range[1] (see below footnote†). The SQS-23 and older systems, with their lesser detection ranges, obviously did not meet the "over-the-horizon" definition.

Three Major Exploitable Paths

A second characteristic of *long-range,* echo-ranging sonar (as defined in figure 1) is the ability to exploit, given the right conditions, one or more of the following three major types of deep-ocean sound paths from a hull-mounted sonar on a surface ship to a submarine target: (1) surface duct, (2) bottom bounce, or (3) convergence zone.

The convergence zone propagation path produces high-intensity zones a few miles wide at multiples of about 30 to 35 miles in the mid-latitudes of the Atlantic and Pacific oceans in water depths greater than

---

*Appendix A, found at the end of this book, contains a chronology of events related to the SQS-26 program. Appendix B provides a listing of some of the personnel associated with SQS-26 development, along with their respective affiliations.

†Endnotes containing source citations and oftentimes additional substantive information are found in chapter 11 of this book.

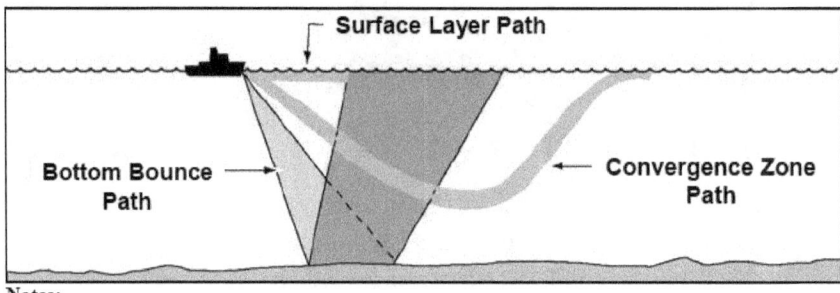

Notes:
1. The vertical scale is exaggerated to show more detail in the paths.
2. The isothermal layer that is typically found within the upper few hundred feet of the ocean forms the surface layer (or surface duct) path. Upward refraction and surface reflection in the surface layer produces a surface duct that at the lower frequencies can provide submarine detections to long ranges — as long as the submarine is in the duct. A major attraction of the bottom bounce and convergence zone paths is that they provide performance that is little affected by submarine depth. Older active sonar systems could exploit only paths within the upper few hundred feet of the ocean.

Figure 1. Deep-Ocean Sound Paths Exploitable by the SQS-26 Long-Range, Echo-Ranging Sonar

about 2000 fathoms. This phenomenon was studied both theoretically and experimentally by Maurice Ewing's wartime research group at Woods Hole.[2]

SCOPE OF SQS-26 HISTORY

The attention here will be largely confined to the "SQS-26 years" (1955 to 1975). The SQS-26 story will be divided into nine parts: (1) historical background, (2) early concept formulation, (3) full-scale experimentation and development, (4) prototype testing, (5) supporting research, development, and testing, (6) the rubber dome window, (7) guidance on equipment operation and tactical employment, (8) Fleet performance, and (9) essential ingredients in program success. These parts are presented in approximate chronological order, although there is some noticeable overlap.

As stated previously, the SQS-26 development period* extended over a period of 20 years from 1955 through 1975, the year in which the

---

*Guidance on equipment operation, tactical employment, and early Fleet observations will be considered here as part of SQS-26 development.

Chapter 1 — Introduction

last SQS-26 ship was commissioned. In September 1972, the Operational Test and Evaluation Force (OPTEVFOR) issued the final operational appraisal of the SQS-26 (CX). It was not until 1973, however, that the rubber dome window, the single most important improvement to the SQS-26 system, was evaluated by the Naval Underwater Systems Center (NUSC) in the presence of the OPTEVFOR observers. Finally, it was 1975 before the analysis of the output of both supporting research programs and production equipment behavior was fully digested. Only then could a useful model of total system performance be completed and operating guidelines provided to the Fleet.

SYSTEM LONGEVITY

The development of the SQS-26 was arguably one of the major tangible accomplishments of the Navy Underwater Sound Laboratory (NUSL) in New London, Connecticut.* Initiated in 1955 entirely within NUSL, the New London organization continued to provide technical direction and scientific support for its subsequent development, production, and upgrade activities over the next 40 years.

---

*In 1970, the Navy Underwater Sound Laboratory (NUSL) in New London, Connecticut, merged with the Naval Underwater Ordnance Station in Newport, Rhode Island, and became the New London Laboratory of the Naval Underwater Systems Center (NUSC). In 1992, it became the New London Detachment of the Naval Undersea Warfare Center (NUWC). Then, in 1996, the New London Detachment was decommissioned and closed, with the New London Laboratory functions and many of its personnel transferred to the Newport Division of NUWC in Rhode Island.

CHAPTER 2
HISTORICAL BACKGROUND

ECHO-RANGING SONAR BEFORE 1955

Chilowsky and Langevin (1914-1918)

Frederick V. Hunt, director of the World War II Harvard Underwater Sound Laboratory (HUSL), has provided the definitive account of the origin of active sonar for submarine detection.[1] In 1914, a Russian (Constantin Chilowsky) made the first proposal regarding the use of acoustic echo-ranging for locating submarines. The idea occurred to him while he was ruminating on the 1912 *Titanic* disaster as he convalesced from tuberculosis in a Swiss mountain hotel. Chilowsky brought his proposal to the attention of the French government, where it ultimately arrived at the desk of distinguished French physicist Paul Langevin in February 1915. Langevin immediately began work on the echo-ranging idea in his Paris laboratory at the School of Industrial Physics and Chemistry.

For a sound source, Langevin devised an electrostatic capacitor that employed a sheet of mica for the dielectric, with the water that was in contact with the mica used for one electrode. The mica mechanically moved in response to the electric field created by a high-frequency electrical transmitter connected to a second insulated electrode. For a receiver, Langevin used a waterproofed carbon microphone. By March 1916, he had achieved one-way acoustic transmission across the Seine. In their last act of collaboration, Langevin and Chilowsky drafted a joint patent application for the submarine echo-ranging concept. At Langevin's request, Chilowsky detached himself from the experimental program after what Hunt tells us was a "less than entirely serene collaboration."

Langevin continued with the research, moving his experimental operations to Toulon in April 1916. With an improved system, he demonstrated the reception of echoes from a large iron plate at 200 meters. By 1918, the system had evolved to the point where echoes could be received from a submarine off Toulon at 1500 meters. In the historical summary by Elias Klein, one of the early sonar designers in the

United States, Langevin's operating frequency in this first echo-ranging equipment was shown to be 40 kilohertz.[2]

Langevin's sound projector, which was used for both transmission and reception, consisted of a "sandwich" of quartz crystal between two steel plates. Quartz exhibits a "piezoelectric" effect that provides a change in the dimensions of the quartz when the electrostatic surface charge is varied. The effect also works in reverse, with compression in the quartz from an incident sound wave changing the surface charge on the quartz. Devices for converting mechanical energy (such as that produced by underwater sound) to electrical energy and vice versa are now called "transducers."

Langevin's sonar was not developed in time to be operationally employed in World War I.

Sonar Development at the Naval Research Laboratory (1923-1941)

Formally established in 1923, the Naval Research Laboratory (NRL) in Washington, DC, continued echo-ranging sonar development, building on the pioneering work of Langevin. Klein, who joined the NRL staff in 1927, tells us that funds for scientific research in the military field were scarce at that time. As a matter of fact, the NRL Sound Division, which constituted the Navy's sole effort on underwater acoustics, numbered only five scientists when Klein was first hired.[3] The first superintendent of the Sound Division was Dr. Harvey C. Hayes, who continued in this capacity throughout World War II.

By 1927, the NRL Sound Division had completed the development of the "QA" sonar, the first destroyer-mounted, echo-ranging sonar in the U.S. Navy.[4] The QA employed a transducer design based on Langevin's 1918 quartz-steel sandwich.[5] According to Ralph DelSanto's history, tests on the QA at a location off Key West, Florida, produced submarine echoes at ranges up to 1 mile.[6] By 1933, the Navy had installed the QA sonar on eight destroyers.[7]

When the United States entered World War II in 1941, the NRL QA design had evolved into more than a dozen types of sonar, beginning with the designation QB or QC to indicate the type of application that was being accommodated.[8] The major contractor was the Submarine Signal Company, later a division of Raytheon.

In the QB sonars, synthetic Rochelle-salt piezoelectric crystals replaced the quartz, and in the QC series, magnetostrictive transducers replaced the Rochelle-salt crystals. The magnetostrictive effect makes use of the change in the length of a ferromagnetic material that is subjected to a magnetic field. The QC series transducer assembly consisted of elements mounted on a 15- to 18-inch steel "banjo" that could be rotated in azimuth about the long axis. Later on, the standard wartime sonar became the QGB, manufactured by RCA. Four hundred of these were delivered to the U.S. Navy in 1944. To detect a submarine at an unknown bearing, the QGB sonar "searched around with beams of sound about 20° wide, with a 'ping-listen' operation on each bearing. About 4 minutes were required to complete a 360° sweep. When a return echo was received, a timing circuit determined range, and the bearing was read by comparing the direction in which the transducer was trained with the ship's heading given by a gyrocompass."[9]

The early World War II sonars were encased in a 19-inch spherical dome, which was later streamlined to a teardrop shape to reduce vibrations and turbulence from the flow of water over the dome face. The typical World War II maximum detection range remained at about 1 mile.

Scanning Sonar Development at the Harvard Underwater Sound Laboratory (1941-1945)

During World War II, HUSL (directed by F. V. Hunt) developed the first *scanning* sonar, based on a design that provided the capability of 360° coverage in azimuth per ping. Although the war ended before this sonar went into production, Friedman describes its development as "a connecting link to all post-war systems."[10] Sangamo Electric Company built the "QHB," which was a production version of the experimental scanning sonar developed by HUSL.

Chapter 2 — Historical Background

Sangamo's Post-War Sonar Production Sonars (1945-1955)

Sangamo's post-World War II QHB sonar used a fixed cylindrical transducer array that consisted of 48 magnetostrictive "staves." The QHB sonar entered the Fleet in 1948.

Figure 2 shows a QHB display of returns of echoes and reverberation. In transmission, all staves were energized in parallel to form an omnidirectional beam in azimuth.[11] In receiving, the staves were rapidly scanned with an electrically phased beamforming network connected to the transducer array with a capacity-coupled rotating switch. At any instant in time, the scanning switch rotor selected 16 transducer elements to feed into a time-delay beamforming network. The receiving beam formed in this fashion was about 10° wide in azimuth. The output of the scanning switch was fed to the input of a receiver. The receiver output was connected to a cathode ray tube

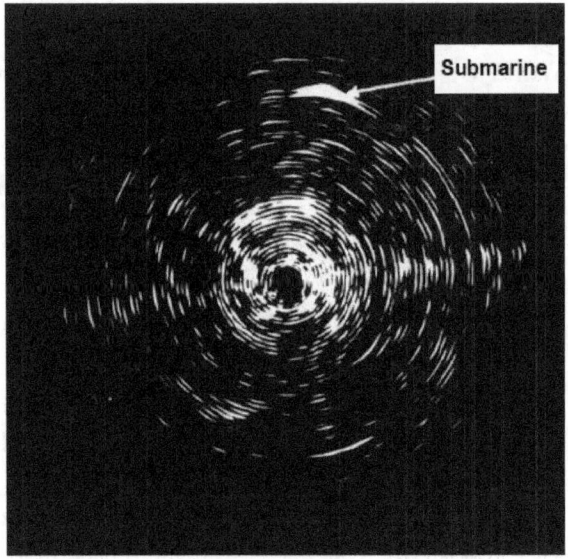

Note: A submarine echo is indicated here at a relative bearing of 20 and a range of about 1.5 kiloyards. The background interference indicates the spiral nature of the cathode ray tube sweep. A new sweep is generated with each transmission; the interval between transmissions here is about 3 seconds.

Figure 2. Display on Scanning Sonar Developed by HUSL During World War II

display, with the electron beam position on the cathode ray tube synchronized with the scanning switch rotation in a manner such that any echo impinging on the array would brighten the display at the correct bearing. The radial displacement of the spirally expanding, cathode-ray-beam sweep was synchronized with the outgoing transmission, elapsed time, and speed of sound. This process was performed such that the displacement from the center of the display of any returning echo would represent range to the corresponding sonar target.[12]

The outgoing pulse duration of the QHB was 35 milliseconds, which meant that the scanning switch rotation rate had to be sufficient to cover a 360° scan in 35 milliseconds. When the QHB sonar entered the Fleet in 1948, its typical sonar range was still limited to about 1 mile. However, its continuous 360° geographic display of all acoustic targets within range was a considerable advance over the older "searchlight" sonars.

A Sangamo version of the QHB became the SQS-10, entering the Fleet in 1950.[13] The next sonar development was Sangamo's SQS-4, a scaled-up version of the SQS-10, which entered the Fleet in 1954.

Other Sangamo models at lower frequencies followed, with the original a version eventually becoming the SQS-4 Mod 4. Before this, a a modification with the same dimensions as the Mod 4 was designated as the Mod 3; a modification with larger dimensions became the Mod 2; and an another modification became the Mod 1. The frequency diversity achieved by the four SQS-4 mods was useful for minimizing mutual sonar interference among ships in a formation.

The SQS-4 was considered to have a typical detection range on a periscope depth target of 5 kiloyards, a sizeable increase over the 1-mile detection ranges* experienced with predecessor systems.[14]

---

*Mile in this context shall mean nautical mile, which is equal to about 2 kiloyards (or, more precisely, 2.025 kiloyards).

Chapter 2 — Historical Background

POST-WAR PERCEPTION OF THE SOVIET SUBMARINE THREAT

Soviet Submarine Technology

Despite the progress that was seen with the introduction of longer range sonars into the Fleet in the decade following World War II, there were reasons for serious concern — primarily over the Soviet Navy's inheritance of a German submarine design technology that was far superior to their own. One such example of German enterprise was the Type XXI submarine with snorkel, which did not have to surface to charge its batteries and could move at 17.5 knots while submerged.

But even more menacing was the Type XXVI "Walther" boat still on the drawing boards. It was to be powered by a hydrogen-peroxide turbine that required no access to the outside atmosphere. Based on experimentation conducted at sea starting in 1940, it was predicted that this submarine would be able to travel completely submerged for 158 miles at 25 knots.[15] Further British experimentation in the 1950's was to reveal that the concept was impractical for long-range, ocean-going submarines.[16] Nevertheless the post-war perceptions of the Walther potential increased concern for how the Soviet Navy might benefit from German submarine technology.

In the late 1940's, Western military leaders feared that the following influences would combine to produce a massive force of highly effective "Red" submarines: (1) the Soviet interest in submarine development, (2) the high priority given to rehabilitating Soviet shipyards, and (3) the available German U-boat technology. In 1948, Rear Admiral Charles B. Momsen stated to the Navy General Board that the Soviets could have as many as 2,000 submarines of all types at sea within 10 years.[17] These submarines were envisioned as having the potential to prevent the United States and her Allies from operating overseas, thus permitting Soviet land forces to overrun Europe. This scenario would cause serious concern to the U.S. Navy for the next four decades.[18]

In 1950, a study concluded that 5 years after World War II the U.S. Navy still lacked the means to counter the German-designed Type XXI boat. In addition, further Soviet advances in propulsion and weaponry were expected regarding nuclear-propelled submarines and submarine-guided missiles capable of carrying atomic warheads. If the U.S. Navy were to meet the potential Soviet threat, *technological development could not be delayed.*[19]

In the early 1950's, the nuclear submarine was to become a reality. USS *Nautilus* had been conceived in 1949 and launched in 1954. Drew Middleton eloquently summed up the reaction to the launching:

> . . . She was 300 feet long, capable of speeds of 25 knots submerged for a period of 50 days. The launching of the *Nautilus* began a new chapter in the history of the submarine, a chapter that is still being written. The moment the *Nautilus* went down the ways . . . the destroyer's old prey became the hunter not only of other submarines but of the destroyers themselves. A new and frightening military age had dawned.[20]

Sensitivity of Sonar Performance to Submarine Depth

Another worry was the sensitivity of the detection performance of the post-war sonars to submarine depth. Despite advances in submarine detection performance on a submarine within the isothermal surface layer, once the submarine went deep, typical detection ranges could not be expected to be significantly greater than the 2-kiloyard capability associated with the echo-ranging sonars of the last three decades.

The reason for the difficulty in detecting a submarine below the surface layer is illustrated in figure 3, which has been excerpted from Urick.[21] By way of explanation, the surface layer of water is typically isothermal but at some point below the surface (in this case 200 feet), the water will begin to become colder and continue to do so as depth increases. Isothermal layer depths typically run from 50 feet in the summer to 400 feet in the winter.

In the figure, the elevation of the sound rays within the isothermal layer starting out at the source is shown, with the negative values indicating the angle below the horizontal as measured at the source. Within the isothermal surface layer, low-elevation sound rays are refracted upward by the increase in sound velocity, caused by the increase in pressure with depth. The upward-refracted rays finally encounter the sea surface, where they are reflected in a forward direction back into the isothermal layer. This behavior produces a favorable

Chapter 2 — Historical Background

ducting effect for the rays shown in this particular example (i.e., the angles are less negative than $-1.76°$).

Also seen in the figure are the rays with elevation angles more negative than $-1.76°$, which are found to reach the downward refraction region where a decrease in sound velocity is produced by the decrease in temperature with depth. This behavior creates a below-layer "shadow zone." The sonar detection range on a below-layer target depends upon the target depth, temperature decrease with depth, depth of the isothermal layer, and reverberation created by favorable propagation to scatterers within the isothermal layer. An SQS-4 detection range of 2 kiloyards was typical for a submarine below the isothermal layer. While lowering frequency and increasing power could substantially increase detection ranges on a submarine within the isothermal layer, these same measures had comparatively little effect on below-layer detection ranges.

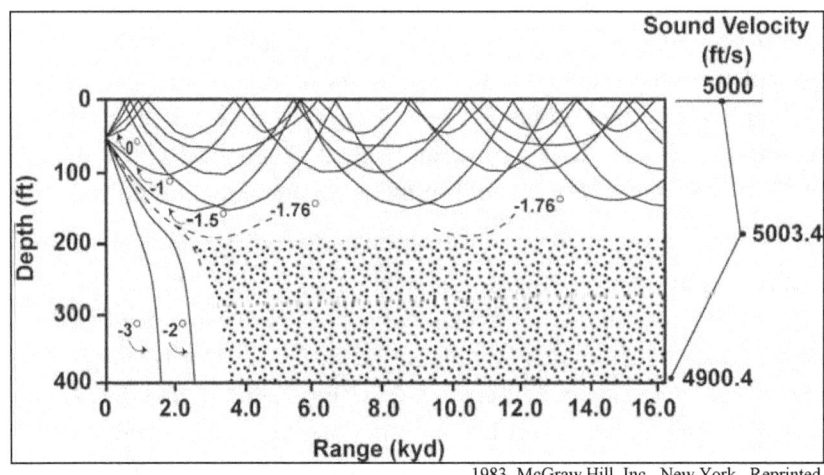

1983, McGraw Hill, Inc., New York Reprinted with permission Taken from a drawing by R Urick in *Principles of Underwater Sound*[21]

Notes:
1. The vertical scale is expanded.
2. The rays from a 50-foot source are labeled by their elevation at the source. The increase in pressure with depth in the isothermal layer produces an increase in sound velocity with depth and consequent upward refraction. For angles less negative than -1.76, the upward refraction and surface reflection produce a favorable ducting effect. Rays with more negative angles escape from the duct into the region where sound velocity decreases with depth as a result of the decrease in temperature with depth, causing a "shadow zone" below the isothermal layer.

Figure 3. The Shadow Zone Below a 200-Foot Isothermal Layer

The above discussion does not mean that the detection range improvements on a target within the isothermal layer were unimportant. The submarine still had a need to enter the isothermal layer to use its periscope and/or its radar for (1) reconnaissance, (2) surface target identification, and (3) range determination. On occasion, the submarine antennas would be used for transmitting or receiving communications in this layer. However, whenever the submarine went below the isothermal layer, its passive listening capability could be seriously degraded. Despite the advantages of the submarine operating in the isothermal layer, it could go below the layer to avoid detection by a surface sonar.

In the 1950's, NUSL experimented with "variable depth sonar" and found that coverage of a deep target could be substantially improved a large percentage of the time when a surface ship towed its sonar at depths in the 300- to 500-foot strata. Long-range refraction sound channels are often found by lowering a sonar array to this region.

However, at latitudes within roughly 30° of the equator, such long-range refraction sound channels generally do not exist and the detection ranges against a deep target are not significantly greater than those obtained from a hull-mounted sonar. This situation is the result of strong negative temperature gradients usually found in these latitudes (more negative than 2° per hundred feet), which seriously limit performance with the variable depth sonar technique. Still another problem surfaces when the active sonar range is extended to the necessary 5 to 10 miles (even where favorable refraction channels exist) because this approach requires an array so large that variable depth operation tends to be impractical.

Memories of World War II

A final contributor to the concern about post-war submarine threats was that most of the U.S. naval officers who had served at sea during World War II now had the responsibility for making key decisions about the future course of the Navy. Within the previous decade, these officers had personally witnessed the horrors of unrestricted submarine warfare. With German submarines responsible for sinking approximately 2,575 Allied and neutral ships, some 45,000 sailors in the Allied forces had been killed, often by drowning.[22]

Chapter 2 — Historical Background

POST-WAR RESEARCH ON LONG-RANGE SOUND PATHS

The Navy's interest in the "technological development" recommended in the aforementioned 1950's study was already evident in the late 1940's. One result was an intensive emphasis by the Navy laboratories on the long-range sound paths that might be exploitable for echo-ranging detection of submarines.

Three post-World War II Navy laboratories were to play a major role in future ASW sonar improvements: (1) the Naval Electronics Laboratory (NEL) in San Diego, California, (2) the Navy Underwater Sound Laboratory (NUSL) in New London, Connecticut, and (3) the Naval Research Laboratory (NRL) in Washington, DC. NEL was an outgrowth of the wartime contract operation at San Diego under the University of California's Division of War Research. NUSL was a post-war combination of two wartime contract operations: (1) the Columbia Division of War Research in New London, Connecticut, and (2) the Harvard Underwater Sound Laboratory (HUSL) in Cambridge, Massachusetts. Both NEL and NUSL were placed under the Navy's Bureau of Ships (BuShips),[23] while NRL continued to perform under the newly organized Office of Naval Research (ONR).

Naval Research Laboratory

In 1948, NRL initiated a program of long-range propagation studies with the objective of extending detection ranges against quiet submarines. The result was a program of echo-ranging sonar research concentrating on (1) near-surface ducting and (2) bottom reflection in deep water.

In 1951, NRL scientist Robert J. Urick, later to write the standard sonar reference work,[24] wrote an article foreseeing the tactical use of the bottom reflection path for submarine detection with echo-ranging sonar.[25] He pointed out that the bottom path ray geometry, with its large inclination angles that are relatively immune to refraction, avoids the shadow zone effect experienced by surface duct sonars. With remarkable insight, he foresaw that the tactical exploitation of bottom-reflected sound would introduce new complications in sonar employment. Performance would depend upon two considerations that had previously been of no importance to sonar effectiveness: (1) knowledge of the reflectivity of the bottom and (2) operator decisions on the appropriate

Chapter 2 — Historical Background

tilt angle for the sonar beam. To this end, the bottom of the ocean would have to be mapped for the quality of its reflectivity to distinguish acceptable from unacceptable locations. The decision process for the proper depression angle of the sonar beam and the associated annular coverage zone would require development. The difficulties foreseen by Urick continued to be serious problems in the tactical applications of bottom bounce echo-ranging over the next half century.

Naval Electronics Laboratory

In 1949, NEL initiated experimentation on convergence zone paths by measuring one-way propagation loss from a sound source to a receiving hydrophone at various depths and ranges.[26] In the NEL test areas off the coast of California, observations were made of high-intensity convergence zones (each several miles wide) formed from deep refracted sound paths converging at multiples of about 30 miles. While there was some loss in intensity by a deep hydrophone compared to that experienced by a shallow hydrophone, the hydrophone depth effect was small compared to that for surface duct propagation. Thus, it seemed that the effect of submarine depth would also be small if this path were to be used for echo-ranging in the detection of submarines.

Navy Underwater Sound Laboratory

In 1949, NUSL initiated the Acoustic, Meteorological, Oceanographic Survey (AMOS) program in cooperation with the Navy Hydrographic Office — a program that would last for 5 years.

During the AMOS effort, the acoustic properties of the sea lanes of vital interest in the North Atlantic were to be experimentally examined on a series of cruises. The output of the AMOS program was to be used to provide predictions of performance for existing sonars and design information for new sonars.

In addition, an extensive program of "figure-of-merit" measurements for current sonars was initiated in the Fleet to permit converting the propagation loss information to detection range predictions. These measurements involved source level, noise level, and minimum detectable signal level versus range. "Figure of merit" was defined as the source level minus the minimum detectable signal, where both terms were described in "decibel" units.* The decibel sum of figure of merit and target strength was equal to the allowable propagation loss for a 50% probability of detection. The range at which the actual propagation loss equaled the allowable loss represented the predicted detection range.

In January of 1954, J. Warren Horton at NUSL, using Morris Schulkin's preliminary analysis of AMOS data on bottom reflection loss versus frequency and angle, made an analytical determination of the optimum frequency, power, and array size required for bottom-reflection echo-ranging at ranges of 40 kiloyards.[27] This effort resulted in the first conceptual design of a long-range active sonar that promised both to meet submarine detection requirements and to fit on current ASW ships.

Figure 4 shows a rare group photograph of the three pioneers of active sonar mentioned in the foregoing account. Pictured are NUSL's J. Warren Horton, HUSL's Frederick V. (Ted) Hunt, and NRL's Harvey C. Hayes, all outstanding scientists and leaders. Taken in 1966 upon the occasion of the 25th anniversary of the New London Underwater Sound Laboratory, the photograph shows (in the foreground) one of the earliest mechanisms constructed for use as an underwater sound source. Built around 1912 by the Submarine Signaling Company (now part of Raytheon), this device could be suspended beneath a light ship to provide an all-weather underwater beacon to passing ship traffic.

FRUITION OF RESEARCH PROGRAMS

The year 1955 yielded an outpouring of information on the prospects of long-range, echo-ranging sonar improvements from the Navy laboratory experimentation and studies initiated in the late 1940's.

---

*The decibel is an engineering term expressing the ratio of two intensities. If r is the ratio of interest, the corresponding decibel (dB) value is 10 log r. If r is 2, the corresponding decibel expression is 3 dB; for r = 4, it is 6 dB; for r = 10, it is 10 dB. Adding decibels is the same as multiplying intensity ratios. Thus, a ratio of 2 (3 dB) times a ratio of 10 (10 dB) is 13 dB; a ratio of 10 (10 dB) times a ratio of 10 (10 dB) is 20 dB. Propagation loss is the ratio of the sound intensity received at a specified distance from the sound source to the sound intensity measured at a unit distance from the source.

Chapter 2 — Historical Background

From left to right, HUSL's Frederick V. Hunt, NUSL's J. Warren Horton, and NRL's Harvey C. Hayes

Figure 4. Three Early Pioneers of Active Sonar

AMOS Data Analysis at NUSL

In March 1955, H. Wysor Marsh and Morris Schulkin published the key results from NUSL's recently completed AMOS program, which showed the first statistical behavior of bottom loss versus frequency and angle as taken from the large body of AMOS data.[28]

Convergence Zone Echo-Ranging by NEL

By early 1955, NEL had installed an experimental echo-ranging system capable of using frequencies below 2000 Hertz on USS *Baya* (AGSS-318). In May, NEL obtained successful echo-ranging against submarine targets in the first convergence zone. Within another year, echoes were obtained from the third convergence zone at a range of 100 nautical miles.* While this demonstration was a long way from providing a 100-mile detection capability on a Fleet ship, it was undeniably a major step in demonstrating progress in what had seemed to be a plateau in echo-ranging detection capability for the previous 30 years. The NEL work produced considerable optimism in the Navy, although there was also a concern about the practicality of exploiting zonal coverage of only a few miles at ranges as far as 100 miles away. Furthermore, the size of the experimental system on *Baya* was such as to preclude installation on operational ASW platforms.

Bottom Reflection and Surface Duct Echo-Ranging by NRL

In August 1955, NRL published a comprehensive report on its extensive investigations of long-range, echo-ranging at 10 kilohertz with an experimental sonar mounted on USS *Guavina* (SSO-362).[29] Studies were made of target strength, signal processing, bottom reflection loss, surface duct propagation, and echo-ranging. This experimentation showed that with the vertical discrimination provided by a tilted vertical sonar beam, bottom reverberation could be separated from echo returns in the near-surface annulus. Echoes were received in water depths up to 2050 fathoms at ranges out to 17.5 kiloyards.

---

*In attendance when the NEL program manager presented these results at the *12th Semiannual Symposium on Underwater Acoustics*, I heard how exciting it had been to witness an improvement by a factor of 100 in the existing 1-mile reliable detection range of the Fleet.

CHAPTER 3

LAUNCHING A LONG-RANGE ACTIVE SONAR PROGRAM: EARLY CONCEPT FORMULATION

CONCEPT FORMULATION STUDIES AT THE NAVY UNDERWATER SOUND LABORATORY

In early 1955, William A. Downes,* head of the NUSL Surface Ship Sonar Department, requested that I conduct a study to determine what should be the next step in sonar development for surface ships. The primary input to this study would be the research information available as a result of the experimental and theoretical studies by NRL, NEL, and NUSL.

Horton's Study

As strange as this may seem, I was unaware of the aforementioned 1954 landmark bottom bounce sonar study by NUSL's Warren Horton when my own study was initiated only a year later. He was not part of the surface ship sonar organization and occupied an office in another building at the New London laboratory. It is possible that his report was not routed to our department or that it remained buried in someone's reading backlog. In any event, this was an advantage because each study could then be seen as an independent investigation of the long-range active sonar problem. In the end, the two separate efforts turned out to be complementary.

Horton used the basic input data from the NUSL AMOS cruise measurements to set up an analytical sonar equation solution that would allow the required array aperture versus frequency to achieve a 50% probability of detection at 40 kiloyards. With differential calculus, he determined the derivative of the array aperture with respect to frequency by setting the derivative of the required area equal to zero, which, in turn, provided the frequency for the minimum required aperture area.

---

*When I first joined NUSL in 1947, I was assigned to work directly for Downes, who at the time was the head of a small group working on a new scanning sonar system design. Although the design was overtaken by events, his guidance as a mentor in my early years at NUSL was of immense value to my development as a design engineer.

Chapter 3 — Early Concept Formulation

The optimum frequency so determined is strongly influenced by the range of interest, but is insensitive to the details of the equipment design. The power was related to area by assuming the cavitation limit of 1 kilowatt of acoustic power per square foot.[1]

Concern with Reverberation

Horton's calculation was based on the assumption that *noise* was the dominant source of interference, although he did recognize the possibility of reverberation masking. Reverberation tends to become more of a problem as range increases because the area of the ocean returning reverberation at any given time is proportional to the product of the pulse length and the width of the chord subtending the angle corresponding to the receiving beamwidth. As range increases, the size of the reverberation area (and thus the reverberation sensed by the sonar) increases proportionally with chord length. The reflectivity of the submarine target, on the other hand, is independent of range.

In the late 1940's, when typical active sonar detection ranges were still limited to 1 mile, I recall Horton remarking that reverberation — which was not considered a serious issue at short range — could be a problem at greater ranges. In 1950, he thought that reverberation might prevent successful echo-ranging even at 10 kiloyards. Subsequent experimentation with the lower frequency scanning sonars in the early 1950's, however, revealed that his concern was unfounded. What would happen at 40 kiloyards, on the other hand, was still another matter. Little was known at that time about quantitative reverberation expectations at such ranges — especially for angles of incidence occurring when the surface was involved with a bottom-reflection path.

Bell Study

Studying what it would take to echo-range via the bottom-reflected path led me to choose a system design frequency that would maximize "echo excess," i.e., the ratio of the actual echo level (in a noise background) to that required for detection. Echo excess can be related to detection probability by Schulkin's method, which was developed in the early 1950's.[2] Thus, I simply plotted the calculated echo excess as a function of frequency; the frequency corresponding to the maximum echo excess was the "optimum frequency." After the optimum frequency was found, I determined the size of the system and the required power to provide echo excess adequate for a reasonable detection probability.

The advantage of echo excess as a measure of effectiveness is that it combines all the ingredients of the sonar equation into a single quantity and permits setting up the sonar equation for a detection probability other than 50%. Rather than use Horton's conventional 50% probability goal to determine system size requirements, I used a 90% criterion because the bottom-reflected sound path in combination with a vertical sonar beam produces an annular coverage region. In contrast to surface duct coverage, which tends to grow stronger as a target is closed, the bottom-reflection annular coverage grows stronger only as the target closes from the outer edge of the coverage annulus to the center. At lesser ranges, the echo excess weakens as the target is closed further. Thus, the 50% detection probability criteria for the center of the annulus would mean that a number of detection failures approaching 50% would occur. Use of the 90% detection probability criterion, of course, resulted in the specification of a larger sonar array than that required for Horton's 50% criterion.

Another difference in the two approaches was that I could determine how much echo excess was lost from the use of a frequency that was different from the optimum one, as shown on the plots of echo excess versus frequency at two ranges in figure 5. This determination was important because the optimum frequency for a 40-kiloyard target was different from that for a 20-kiloyard target.

It turned out that the echo-excess loss experienced when using a non-optimum frequency for the target range was much less in the case of a 40-kiloyard optimum frequency than in the case of a 20-kiloyard optimum frequency. This advantage held for all ranges Thus, employment of the 40-kiloyard optimum frequency for system design promised performance that would be less sensitive to target range than using frequencies optimized for shorter ranges.[3]

Chapter 3 — Early Concept Formulation

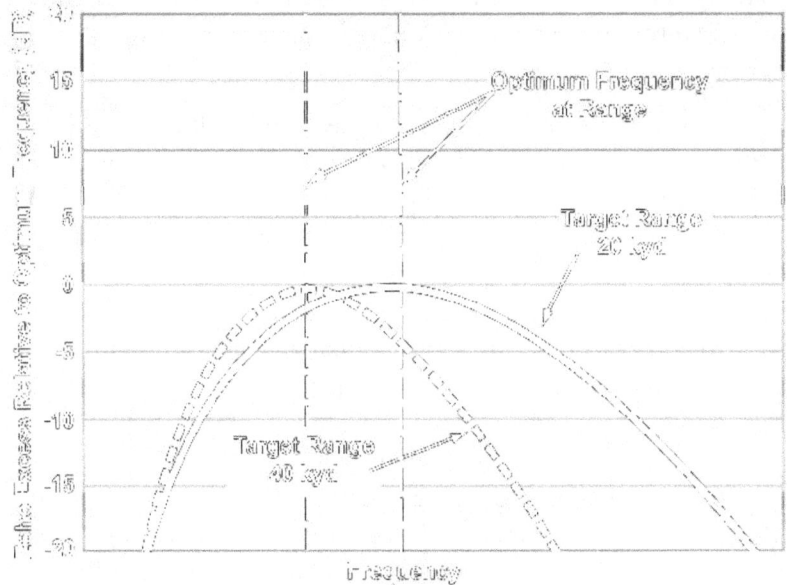

Note: Echo excess is the excess in the echo level over that required for detection. For a background dominated by noise, performance at the higher frequencies decreases as frequency increases because of the increase in attenuation from sea water and the boundaries. Below the "optimum frequency," the increase in noise and beamwidth as frequency decreases overcomes the attenuation effect. Reverberation is not considered in this plot.

Figure 5. Echo Excess Versus Frequency at Two Target Ranges

Horton first introduced the optimum frequency concept in his 1957 book based on lectures to officers of the U.S. Navy; these lectures were part of a naval electronics course set up at Massachusetts Institute of Technology (MIT) after World War II.[4] That an optimum frequency tends to exist for both active and passive sonars is a concept not widely understood, even at this writing some 50 years after the idea was first introduced. A popular misconception is that because propagation loss tends to become smaller with a decrease in frequency, the sonar performance will monotonically improve as frequency is lowered. In practice, at any given range, a frequency will be encountered below which a decrease in performance will occur as a result of the increase in noise and decrease in array directivity at the lower frequencies. Because the propagation loss dependence on frequency is range dependent, the optimum frequency will be a function of target range.

Chapter 3 — Early Concept Formulation

Accommodating a Large Array on a Surface Ship

Although it was initially thought that the solution might be a towed "billboard" array, it was found that the coverage provided by this type of array in the important forward sector was degraded by the impracticality of electrically steering a beam parallel to the array face.

There was no simple way to determine how large a *hull-mounted* array could be accommodated on a surface ship. It should also be noted at this point that the ship designer always wishes to keep the array size to a minimum because of its impact on ship cost and characteristics.

Harold Nash, head of the Sonar System Department at NUSL (also Downes' supervisor in the organizational hierarchy and a HUSL alumnus) cautioned us not to adopt the "21-inch" mentality in array design. This limitation on array size was the result of a constraint related to the available space between ship frames.

In the early 1940's, the Navy had decided on a standard 21-inch sonar fitting between destroyer frames to accommodate a hoist for bringing the sonar inboard for inspection, repairs, or replacement. The HUSL discussion of sonar design considerations in "Scanning Sonar Systems"[5] stated the following: "The diameter of the transducer is determined in part by the size of the opening through which it must pass to be let out of the bottom of the ship." This report also refers to a sketch of the 21-inch hull penetration for the contemporary QC trainable "searchlight" sonar. The QC flange was initially used to mount the experimental scanning sonar cylindrical arrays, which were limited by the QC fitting.

While the constraints seem reasonable for initial experimentation, the HUSL discussion of "future developments" contained no mention of proposals to increase array size. This situation seems strange in retrospect since NUSL, in later years, tended to follow the guideline that states "as the array goes, so goes the sonar."[6] In defense of HUSL, the priority during its 5-year existence was the attempt to make scanning sonar a practical realization, which had been accomplished quite well by the end of the war. By the time the larger arrays were developed and began to appear in the surface ship fleet, the wartime HUSL organization had been closed for a decade.

Chapter 3 — Early Concept Formulation

To acquire some feeling for what the constraints might be in providing a aperture, NUSL mechanical engineer Walter Whitaker and I looked at the plans of the existing *Dealey* class destroyer escort (DE), a candidate platform. The 1817-ton *Dealey*, laid down in 1952, was the first post-World War II ship built for convoy escort duty. Examining the forward cross section of the ship, we sketched a dome at the conventional location that would provide a streamlined enclosure for an array. The initial vertical dimension was determined from an examination of NRL plots of reverberation versus time for a bottom bounce echo-ranging geometry, and it appeared adequate to "resolve out" the bottom reverberation from the expected echo return while forming a reasonably wide near-surface detection annulus.

Whitaker next installed a scale-model version of such a dome on an existing 6-foot-long wooden model of the *Dealey*. Satisfied that it looked appropriate, the proposal was then made that such a dome could be used to provide a streamlined enclosure for the initially conceived flat-face array. This array would be mechanically trainable in both the vertical and azimuthal plane.

After the sonar array design had been "locked in", NUSL later discovered from David Taylor Model Basin experts that minimizing drag meant mounting the array and dome at the very front end of the ship, much as bulbous bows are added to large tankers to reduce their drag resistance. The required dome size in that position, however, no longer looked visually reasonable for a small escort ship. Flared out from the stem of the ship, the dome lateral dimension viewed head on seemed to be an ungainly appendage. Had it been known earlier that the sonar would have to be mounted in the bow location, some rethinking about the NUSL size requirement might have occurred. However, the ship designers seemed to take all this in stride, making no objection to providing such a large dome at the bow. As it turned out, the maximum speed of an escort ship was not affected by the bow dome, but the *cruising speed* was diminished somewhat because of the resistance that the dome caused.

In later years, Commander Clark Graham discussed the impact of sonar equipment of this size from a ship designer's perspective.[7] Considering both the direct and indirect effects of (1) added drag at cruising speed, (2) sonar weight, (3) extra space for sonar personnel, and (4) sonar power requirements, he concluded that the addition of a sonar array of the size

NUSL had specified, along with the associated equipment and personnel, would add some 600 tons to the displacement of a modern destroyer.

INTERACTIONS WITH THE OUTSIDE WORLD
Scout Ship Concept

By early summer of 1955, the basic study was completed. On 27 July, I visited the Office of the Chief of Naval Operations (CNO) in Washington, DC, to discuss requirements and NRL to discuss their bottom reflection echo-ranging results.

At CNO, I visited Commander Leslie J. O'Brien, who was working on a self-initiated study of a sonar ship. This ASW ship, which would be optimized for carrying a sonar system,* was later referred to as a "scout ship." The premises of the scout ship concept were as follows:

- Convoy escort ships are ineffective against the most modern high-speed U.S. submarines.

- It must be assumed that in a relatively short time enemy submarines will exist that will equal the capabilities of the best U.S. submarines.

- ASW weapons and sonar must be greatly improved to counter this threat.

- The preceding premise means that weapons and sonars, in all probability, will be heavier.

- If both weapons and sonar are to be mounted in a single ship, the ship will be large and expensive and will not be optimum for either system.

- Convoy protection can probably be handled with minimum cost and maximum effectiveness as a team effort by sonar ships and weapons carriers (the latter to include both aircraft and ships).

---

*During World War II, O'Brien was decorated for "courage and devotion to duty under fire." He was Executive Officer of USS *Van Valkenburgh*, one of the destroyers assigned to radar picket duty during the 1945 Okinawa campaign. The ship successfully fought off Japanese suicide attacks by destroying four aircraft, assisting in the destruction of three more, and driving off all other attackers. In 1965, he was promoted to rear admiral, and in 1967, he became Director of the Antisubmarine Warfare and Ocean Surveillance Division at CNO.

Chapter 3 — Early Concept Formulation

The characteristics of the sonar ship were arrived at through the following logic:

- The enemy submarine will begin to be dangerous about 5 to 10 miles away from the sonar ship.

- The preceding scenario requires a detection range of about 5 to 10 miles.

- Appreciably longer detection ranges would be of marginal advantage because of weapons system limitations.

- Even if longer range sonar capabilities were readily available, it might be better to concentrate on a thorough coverage of the 5- to 10-mile region.

- While a submarine would seem to be an ideal sonar platform, it would be more expensive than a surface ship and the training problem would be more difficult.

- The first sonar ship would be built not as a prototype but as a purely experimental ship to obtain design information for future guidance in the development of a prototype. This platform would also be available for expanded sonar experimentation.

- As of this date, no design details are worked out. If the general concept receives high-level approval, a more detailed feasibility study will begin.

Commander O'Brien's view was that the crucial question concerned the inclusion of a weapon on the "sonar ship." Although there was some feeling within CNO that a weapon should be included, O'Brien felt that this approach would tend to control the ship design and that the advantages foreseen for the sonar ship would largely disappear.

Commander O'Brien was initially thinking of a 300-foot ship that had a displacement of 1000 tons, with the sonar 70 feet back from the bow, drawing a depth of 30 feet. Top speed needed to be only 21 knots, in the context of the sonar ship performing only the detection function and weapon delivery being carried out by another ship. I was encouraged to hear that he felt the results of my study would offer support for his project.[8]

Chapter 3 — Early Concept Formulation

Discussions with NRL and BuShips

On 27 July, I visited Dr. Harold Saxton, the Superintendent of the Sound Division at NRL, who was discouraged by NRL's marginal performance in bottom reflection echo-ranging experiments at 10 kilohertz. He expressed an interest in my required system characteristics for a bottom reflection echo-ranging system[9] and thought they showed promise.* At that time, NRL was hoping to become involved in lower frequency, bottom reflection echo-ranging experimentation for the future.

In September 1955, I formally presented my bottom bounce sonar study results at NRL to an audience of both NRL and BuShips representatives. This was the first official presentation to NUSL's Washington sponsors in BuShips.[10]

USAG Symposium Paper

In November 1955, the semiannual symposium sponsored by the Underwater Sound Advisory Group (USAG) was scheduled. Earlier, Downes had proposed that I present my study results on the proposal for bottom bounce sonar. However, I had read in the preliminary material for the meeting that only *experimental* work was of interest (not *proposals*), which was an apparent reaction to the large number of poor-quality proposals delivered at previous symposia. Downes, who had very definite ideas about what he wanted to accomplish, said that these instructions were to be ignored.

Although somewhat apprehensive, I prepared a paper entitled "Fundamental Design Considerations for a Reliable, Long-Range, Echo-Ranging Sonar," which was to be delivered on 8 November 1955 at the *12th Navy Symposium of Underwater Acoustics* held at the University of Pennsylvania. After arrival at the symposium, I found it unsettling to first listen to an impressive *experimental* paper by NEL on first, second, and third convergence zone echo-ranging experiments. However, the system that I was discussing would also be able to exploit surface duct,

---

*Although Dr. Saxton was some 20 years my senior, I was impressed with the consideration that he showed me during my visit. At the end of the discussion, he found that I had taken the bus out to NRL, whereupon he escorted me to his car and personally drove me to the bus stop outside the NRL gate, some distance from his office.

Chapter 3 — Early Concept Formulation

bottom bounce, and first convergence zone transmission paths when available. My paper was surprisingly well received, despite the lack of experimental data.

The NEL paper on long-range active detection (LORAD), convergence zone echo-ranging used a frequency of about 800 Hertz. In the discussion period following the presentation of my paper, Harold Nash asked a question concerning the discrepancy between my "optimum" convergence zone frequency and the NEL LORAD frequency. The question was clearly being asked to elicit a reaction from NEL personnel. Although I did not speak of it, I suspected that NEL investigators had not really attempted to determine the optimum frequency.

Although Harold Nash's question provoked no immediate reaction from NEL personnel during the discussion period, the manager of the NEL LORAD program came over to me after the late afternoon session, claiming that I had "planted" the frequency question in the audience. He was extremely offended by what he considered to be a negative interpretation of the way the LORAD program was being conducted.

As it turned out, this event actually initiated an important chain of events on the subject of optimum frequency and the significance of knowing the attenuation coefficient on that frequency. Subsequent to the symposium, I received visits in New London from the NEL director of research, Dr. Gilbert Curl, and then later from its signal processing expert, Dr. James Stewart. Five years later, in 1961, Stewart *et al.*[11] published a paper on the subject of optimum frequency, which referred to my optimum frequency studies. Twenty years later, they published another paper on the same subject.[12]

In a February 2001 telephone conversation, Dr. Stewart revealed that the decision to go to a 1500-Hertz frequency was actually not the result of any optimum frequency studies. The real consideration was that the 800-Hertz frequency allowed too much interference from sea mounts that were located beyond the range that NEL was attempting to cover for submarine detection. As a result, it was concluded that NEL should use a frequency high enough to produce enough attenuation to exclude reception of distant sea mount echoes.

Optimum Frequency and Attenuation

Stewart's 1961 paper dealt with the influence of changes in the understanding of attenuation behavior on optimum frequency. Nash, by his question on my 1955 symposium paper, had played a key role in stimulating NEL interest in the calculation of optimum frequency. He also acted as a catalyst at NUSL in initiating attenuation studies.

At NUSL, an independent analysis by William Wardle called attention to the uncertainty in the attenuation values below 5000 Hertz and the consequent impact on optimum frequency calculations for sonar. Nash became so concerned about poorly understood attenuation that he initiated a NUSL experimental study on the subject. William H. Thorp was assigned as the key investigator and within a few years had completed a seminal piece of experimentation and analysis on the subject of the frequency dependence of long-range attenuation in the ocean.[13] His research was conducted only in the Atlantic Ocean and offered no explanation for the difference in attenuation that NEL had observed in the Pacific. The dependence of attenuation on location was finally definitively solved by the investigations of Mellen *et al.* carried out during the next two decades.[14]

Despite the acrimonious opening in 1955 of a NUSL-NEL dialogue on optimum frequency, the follow-on exchange turned out to be most productive. During the SQS-26 development, I had many stimulating discussions with NEL scientists Kenneth Mackenzie, Melvin Pederson, Frank Hale, Edwin Hamilton, Ernest Anderson, William Batzler, James Stewart, and Eric Barham — all leaders in their specialties. Even though not formally associated with the program, these investigators contributed much valuable information to SQS-26 development.

Melding the NUSL Sonar and CNO Scout Ship Concepts

NUSL was invited to attend a scout ship meeting at BuShips on 14 November 1955.[15] The scout ship project had now evolved from Commander O'Brien's informal proposal for a "sonar ship," which I had encountered 4 months earlier, to a formal BuShips study project. BuShips Code 420 (Preliminary Ship Design) was assigned the main responsibility, with support from BuShips Code 848A (Surface Ship Sonar) and the David Taylor Model Basin.

Chapter 3 — Early Concept Formulation

With regard to what sonar should be used for the project, neither the sonar nor ship design organizations in BuShips were certain that they agreed with the NUSL size requirement. Despite these initial disagreements, all concerned had at least accepted the idea that a large sonar array of some sort should be provided.

BuShips Code 420 showed us results from investigations of six types of ships that could accommodate such a sonar array. One ship had the type of bulbous bow that was finally selected, but another had the sonar array in the conventional location well aft of the bow. Commander O'Brien's influence was evident in the specifications for a 30-foot draft, a 21-knot cruising speed, and a displacement of 1000 tons.

The disappointing reception from BuShips regarding NUSL's sonar requirements was to some extent caused by a lack of formal documentation to back up the verbal presentations on the sonar concept. In December 1955, BuShips requested that NUSL prepare a specification on the sonar system. This task was assigned by Downes to Frank White, who, in a matter of a few days, wrote up a specification based on my initial concept of a trainable flat-face system.[16]

The specification — describing a sonar that could use surface duct, bottom bounce, or convergence zone paths — was internally circulated, discussed, and revised, with John Snow and Harold Morrison making key contributions. A major addition to the conceptual system was the provision for Doppler filtering to improve moving target performance against reverberation, which had been a matter of growing concern. In January 1956, this specification was formally forwarded to BuShips.

The informal decision-making process that Downes adopted during his tenure was to continue throughout SQS-26 development at NUSL. He would schedule meetings of a core group in his organization to hammer out the response to any pending issue requiring action. The group initially consisted of Frank White, Harold Morrison, Russell Baline, Stanley Peterson, John Snow, and myself. If any documentation was required, Downes would usually assign the task to one of the meeting participants.

In arriving at decisions, Downes preferred a general consensus, but, as a man of conviction, he would on occasion continue the debate for extended periods of time as he quietly, yet persistently, attempted to sway us to his point of view. Overall, Downes was an outstanding leader who exhibited common sense, showed great respect for others' views, and managed to elicit the best from his staff.

Although a presentation to the CNO Plans and Policy Group was scheduled for 7 February 1956, a second scout ship meeting was held earlier at BuShips on 17 January.[17] The meeting was called by Commander John F. Kalina of BuShips Code 420 (Preliminary Ship Design) in response to pressure from CNO for a preliminary output in January on the scout ship design. Robert Priest and Philip Mandel from Code 420 were also in attendance, as well as Wesley F. Curtis representing the David Taylor Model Basin.

At this point, the original six ship-design options had been reduced to three. The first, a conventional hull with a bulbous bow, could be included in the FY58 building program. The second, a less conventional hull, would take longer to develop. The third was essentially a fixed-depth submarine with a snorkel tube for air intake.

In accordance with inputs from William Hanley and Elmer Landers of BuShips (Surface Ship Sonar, Code 848A), the depth was specified as 10 feet for reasons that were not clear. The array was to be cylindrical, with steering provided by a combination of mechanical and electrical switching aboard ship.

I raised the point that the sonar size that NUSL felt adequate might be accommodated by a current destroyer escort design and that this possibility should be investigated.

Operational Requirement Problem

At this time, BuShips Code 848A was obtaining inputs not only from NUSL, but also from NRL, NEL, and Sangamo. Although it was understandable from the BuShips perspective why these other organizations should have a voice, the situation was creating a serious problem that was later resolved by the formation of an interlaboratory committee.

An immediate obstacle to adopting the NUSL approach was that Code 848A was quoting CNO guidance stating that the "the recommended system should be based on known techniques" (even though CNO expected major performance improvements). In addition, Code 848A informed NUSL that there was no stated requirement to provide a sonar that could cover ranges out to 20 kiloyards via the bottom bounce path. All this information was, of course, contrary to what I had heard from my discussions at CNO. Unfortunately, up to this point, no representative from CNO had participated in the BuShips/NUSL meetings.

SQS-23 Issue

In 1956, Sangamo was developing a version of the SQS-4, which would eventually be known as the SQS-23. This event no doubt had a strong influence on the BuShips Code 848A proposal that the first scout ship sonar should be a cylindrical array. While a SQS-23-like system would have minimized development problems, it would not have a bottom bounce or convergence zone capability. For an experimental ship that was to be built to accommodate a sonar much larger than predecessor systems, adopting a surface duct-only system seemed (to NUSL) to be incompatible with the philosophy of the scout ship project.

During the weeks following the meeting, BuShips Code 848A, with the help of inputs from Sangamo, countered the NUSL objections by developing a proposal stating that the SQS-23 array could be used to cover *both* bottom reflection and surface duct paths. During the evaluation of the SQS-23, Sangamo participated in testing on the Blake Plateau (east of Florida) in 300 fathoms, indicating that the SQS-23 could make use of bottom reflection paths to reach 10 kiloyards in that particular environment. The existing negative thermal gradient bent the horizontal sound beam downward enough to provide a bottom-reflected path focus zone. While this was a long way from demonstrating a bottom reflection capability in 2500 fathoms, it kept alive the idea of using the SQS-23 for the scout ship sonar.

Code 848A next presented the added argument that the SQS-23 array would fit on currently designed ships with an ASW escort mission. They also noted that NUSL had calculated in its initial study that the SQS-23 used a preferred frequency for the goal of achieving sonar coverage at 20 kiloyards. I then prepared a rebuttal paper describing the objection to the concept of employing an SQS-23-type array on the scout ship.[18]

Early in 1958, the SQS-23 became operational. Ultimately, the Navy produced 197 of these sonars.[19] Once the SQS-23 was accepted for operational use, the pressure continued to employ it as a bottom bounce system in place of NUSL's proposed system, now well on its way to general acceptance. Sangamo experimentally provided a capability on one SQS-23 unit to depress the beams, demonstrating that bottom bounce would be feasible. As a result, I prepared another paper in April 1958 that compared the SQS-23 capabilities as a bottom bounce sonar to those of the NUSL-proposed system. The paper concluded that the SQS-23 array size was insufficient and that the frequency was too high for acceptable bottom bounce performance.[20]

Sangamo was a formidable adversary because of its excellent reputation for building well-engineered, reliable sonars. It turns out that the transducer design indeed involved a considerable amount of development effort on the part of SQS-26 contractors. In 1969, Theodore E. Thuma of General Electric (GE) prepared a summary discussion of nine problems encountered in the effort to produce a reliable SQS-26 transducer element in production quantities. The problems were categorized as (1) cable and connector failures, (2) low insulation resistance, (3) voltage breakdown across the ceramic or internal wiring, (4) excessive changes in capacity or impedance of the elements, (5) leaks and water permeation, (6) parting of the cement joints, (7) deterioration of the head mass encapsulation, (8) transformer failures, and (9) fracture of the ceramic.

Ultimately, the new GE barium titanate ceramic element proved to be more satisfactory in efficiency than the magnetostrictive elements employed on the Sangamo predecessor sonars, despite the aforementioned problems that had to be overcome along the way.[21] Probably more important than transducer problems in the failure of Sangamo to bid on the SQS-26 was the company's unyielding stance that a modified SQS-23 would be adequate to meet future requirements.

Presentation to the CNO Plans and Policy Group

At the 7 February 1956 meeting of the CNO Plans and Policy Group, NUSL's Stanley Peterson presented the material that I had given at the Underwater Acoustics Symposium the previous November directly to the CNO decision-makers. BuShips Code 420 (Preliminary Ship Design) also presented its ship platform study. The NUSL presentation was well received, with the CNO group generally accepting that the scout ship sonar should be capable of operating via the bottom reflection path.[22] At this meeting, representatives of the BuShips Surface Ship Sonar code did not repeat their previous objection concerning the lack of a formal requirement for such a system.

The following month, in March 1956, I finally formally documented NUSL's sonar proposal in a report containing the paper that I had presented in November 1955 at the *12th Symposium on Underwater Acoustics* at the University of Pennsylvania.[23]

Conversion to a Cylindrical Array Concept

In early 1956, NUSL's Downes was beginning to believe that a fixed cylindrical array would be more versatile for the scout ship application than the NUSL-proposed steerable billboard array. The beams would be electrically phased to steer the receiving and transmitting beams in azimuth and in depression, which would permit the option of rapidly switching the beam direction to provide simultaneous surface duct and bottom bounce coverage, with both passive and active reception. A conventional scanning sonar mode for 360° coverage of the surface duct could be provided.

Some of us had reservations about the use of a fixed cylindrical array for bottom reflection path coverage as little was known about the feasibility of electrically steering a cylindrical array to cover the depression angles of interest. There was concern about interactions among the elements that might cause phase and amplitude errors and consequent beamforming problems. On the other hand, installing a mechanically steerable flat-face array on the underside of a surface ship could create a reliability problem in the event of a steering mechanism failure in such an inaccessible location.

Downes also felt that NUSL would inevitably be involved in the testing of much more than an experimental system. If results were favorable, there would be pressure to convert the system into a prototype, which meant that NUSL had better be certain that the essential features of interest for a production system were included. Such features would likely require a cylindrical array.

The outcome resulted in Downes assigning the task of investigating beamforming with a cylindrical array to Harold Morrison. Because of the loss in effective aperture due to the cylindrical array shape and electrical steering, Morrison increased the dimensions of the cylinder, which preserved the horizontal and vertical beam shapes.[24] These dimensions were to become the official requirement for the cylindrical array.

BuShips Acceptance of NUSL Cylindrical Array

By July 1956, BuShips had accepted the cylindrical array for the scout ship.

Although the original study had proposed only a bottom bounce capability, Downes had later asked me to investigate providing a convergence zone capability as well, believing strongly that the system would then attract more support from NEL. Along with most others at NUSL, I was not convinced that the convergence zone capability would be a practical technique. The narrowness of the zone, the large gap between own ship and the zone, the question of providing an attack capability at that range, and the inability to change the zone range to close a submarine that had been detected were all serious concerns.

In the end, it was fortunate that Downes had urged the incorporation of the convergence zone capability in the initial design, because once the Navy was able to experiment with ASW tactics using helicopter or fixed wing support against both exercise and Soviet submarines, NUSL became convinced that the convergence zone capability was important. As it turned out, the frequency, power, array dimensions, signal processing, and displays were just about what was needed to provide performance on both paths. And while there were disadvantages, the convergence zone coverage did not involve bottom reflectivity uncertainties and complex problems in providing a practical means of selecting the right depression angle and zone window coverage.

In the early 1970's, NUSL discovered that the background noise with steel domes did not allow reliable SQS-26 convergence zone performance beyond about 30 miles under free-play conditions. This problem was a concern in North Atlantic operating areas where convergence zone ranges between 35 and 40 miles were common. Before this limitation was known, there were unsuccessful attempts in

Chapter 3 — Early Concept Formulation

the North Atlantic to demonstrate the capability that had been successfully proven in Pacific 28-mile zones and Mediterranean 20-mile zones.

From 1970 to 1971, NUSL worked with Rear Admiral Thomas R. Weschler to incorporate convergence zone detection and tracking tactics into the ASW exercises that he was conducting in the North Atlantic as commander of Cruiser Destroyer Flotilla Two. Although he provided NUSL with every opportunity possible to prove its case, the 35-mile ranges in the operating area were (unknowingly to us at that time) simply beyond current capabilities. With the introduction of the rubber dome window after 1973, however, strong convergence zone performance out to 40 miles was demonstrated.

CNO had requested BuShips to proceed with a conventional ship design for a scout ship that CNO hoped to include in the FY59 building program,[25] with the fiscal year beginning in July 1958. BuShips was to coordinate inputs from NUSL, NEL, NRL, and David Taylor Model Basin on required space and equipment for the sonar. BuShips Code 420 (Preliminary Ship Design) had already designed a dome to accommodate the array.

BRASS EXPERIMENTS

Genesis of the BRASS System

While the foregoing efforts were underway in the NUSL Surface Ship Sonar Division, the Submarine Sonar Division, under the direction of Walter Clearwaters, was laying plans for a way to conduct early experimentation at sea without having to first procure a full-scale system. Because both the surface ship and submarine divisions at NUSL were interested in long-range, echo-ranging sonar development, the two organizations had worked cooperatively on such research since mid-1955.

In 1955, Russell Lewis, Head of NUSL's Exploratory Development Branch in the Submarine Sonar Division, initiated a study of the development of a submarine-mounted experimental sonar and had then traveled to NRL to obtain its view of such a program. NRL, which had fielded a similar submarine-mounted system for its experiments with bottom-reflected paths, thought Lewis's proposal to be a reasonable one.

Chapter 3 — Early Concept Formulation

For experimentation on the physics of bottom reflection echo-ranging, the submarine was a more practical platform than was a surface ship. The deck-mounted experimental array was readily accessible for installation, modification, inspection, and repairs without the need for drydocking. While its capability would initially be modest, the Bottom-Reflected Active Sonar System (BRASS) — so named by Lewis — would permit echo-ranging tests in limited water depths and at relatively short ranges to verify the basic physics of bottom bounce path performance.

BRASS Sea Tests

Hugo Wilms, an electrical engineer in Lewis' branch, was assigned as leader of the experimental program. In January 1956, Wilms began formulating plans for assembling and installing a simple system on the submarine USS *Blenny*. He scheduled a March sea test using a research ship with echo repeaters to simulate a target.[26]

In April 1956, Wilms documented the results of the echo-repeater tests after having explored depths from 100 to 2800 fathoms, frequencies from 1 to 8 kilohertz, and ranges from 0.8 to 40 kiloyards. Under poor surface duct conditions, the bottom-reflected path provided good two-way propagation and "echo" reception with the echo-repeater target, as expected. Wilms was also able to look at the frequency dependence of the bottom-reflected path with his broadband system.[27]

In September 1956, Wilms went to sea again with BRASS to perform echo-ranging tests against an actual submarine target in 100 to 800 fathoms, about 70 miles south of Montauk Point off Long Island, New York. These tests, even though scaled down to relatively short range and shallower water depths, were of considerable interest to CNO because they offered the first solid evidence that bottom bounce echo-ranging on a real submarine target was feasible. The surface ship branch in CNO had been awaiting these results before releasing funds for the more ambitious surface ship experimental system.[28]

In November 1956, a preliminary report issued on the BRASS echo-ranging results indicated that bottom bounce echoes were obtained out to 5.9 kiloyards when no surface duct path was available.[29] A major conclusion of the follow-up analysis was that the bottom loss was in good agreement with the "revised version" of the AMOS model that had been used in my original calculations. Errors in the original AMOS bottom loss figures that required a revision of the bottom loss model will be discussed shortly.

BRASS II

In March 1957, Lewis proposed building a higher capability "BRASS II" experimental system for extending the shallow-water scale model tests conducted in September 1956 into deep water.[30] Figure 6 shows the resulting submarine-mounted transducer and reflector assembly to permit both horizontal and vertical beamforming. The cost of this effort was estimated at $70,000.

Lewis later documented this effort with the preparation of a system block diagram of BRASS II, along with a summary status report.[31]

Deep-Water Bottom Bounce Echo-Ranging

In October 1959, Wilms went to sea with BRASS II to investigate deep-water bottom bounce echo-ranging at 4300 Hertz in locations between the East Coast of the United States and Bermuda. On 1 December, he published a preliminary report[32] stating that his main interest was the 20-kiloyard range. It was found that bottom loss values varied considerably with location, running from 9 dB *above the revised AMOS model* to 12 dB below. (The revised AMOS model is discussed on pages 40 and 41). Wilms observed no problems with reverberation.

Also, no decrease in performance was found when the target or own submarine was below the isothermal layer — a condition that precluded conventional surface duct echo-ranging at 20 kiloyards. All testing was conducted against a beam aspect target.

For the beam aspect submarine target, Wilms found that he could resolve surface reflections from direct paths to obtain target depth information at ranges of 20 kiloyards. At the time, this was regarded as an outstanding accomplishment. However, it was later found that for other target aspects the echo extent was usually comparable to the path difference. This result created an ambiguity between (1) multiple paths produced by the extended target and (2) multiple paths in the medium, thus precluding a resolution of the surface reflected from the direct bottom bounce paths for most operational situations.

Chapter 3 — Early Concept Formulation

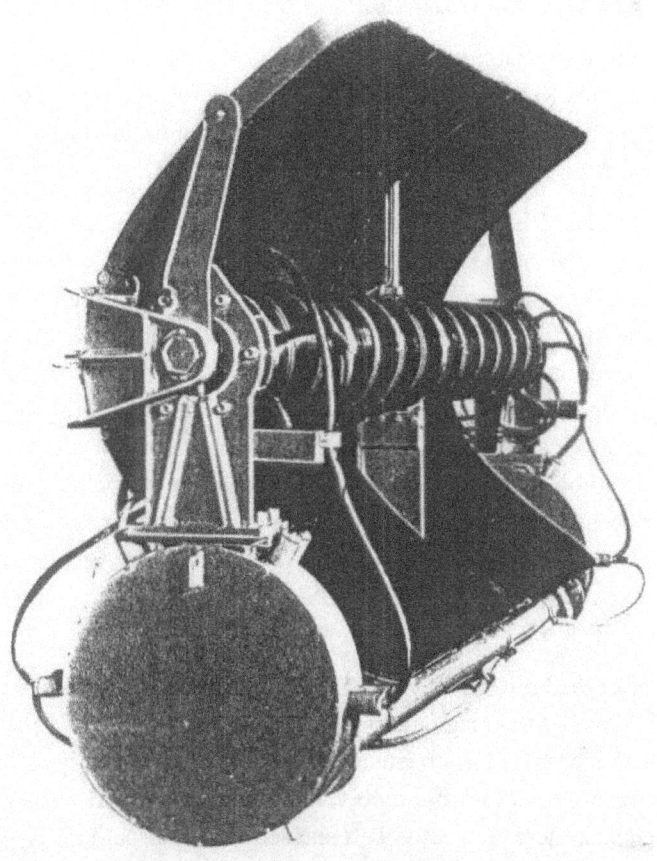

Note: The beam was steered in the vertical plane by rotating a reflector about the horizontal axis of a cylindrical line transducer. Steering in azimuth was accomplished by rotating the whole assembly about its vertical axis.

Figure 6. BRASS II Experimental Array

On 23 December 1960, Wilms presented a summary of bottom loss for BRASS II locations between Bermuda and the East Coast of the United States at the *18th USAG Symposium on Underwater Acoustics* held in Monterey, California. This discussion, which provided the first information on the variability of bottom loss with location, was the beginning of bottom loss charting that would define regions of good and poor performance for the bottom bounce path.

Chapter 3 — Early Concept Formulation

Discovery of Errors in AMOS Bottom Loss Values

The revised version of the measured AMOS bottom loss referred to by Wilms in his May 1957 report showed higher losses than the originally published curves from the AMOS program analysis, which requires an explanation.

Lloyd "Ted" Einstein of the Submarine Sonar Branch at NUSL, a pioneer in the use of the central NUSL computer for sonar modeling, had been assigned the task of making the calculations regarding the expected performance for submarine active bottom bounce sonar. In the process, he questioned me concerning the surface-reflected path contribution to two-way propagation loss. While the depressed-beam geometry would involve no surface reflection path near the sonar, there would be surface reflections involved in the vicinity of the target for the paths going both to and from the target — reflections that would tend to reduce the propagation loss. Einstein wondered if these surface-reflected paths had been considered in the original AMOS program analysis of bottom path measurements.

We discussed the surface-reflected path issue with Morris Schulkin, who had provided the original curves[33] and who gave us the sobering news that this path had *not* been considered in the data analysis. For the nondirectional projectors and receivers used in the one-way AMOS measurements, there was actually a surface-reflected path in existence near *both* the projector and the receiver. The 6-dB decrease in propagation loss via the bottom from these surface reflection paths had been implicitly subtracted from the real bottom loss by Schulkin's process of ignoring these paths, thus giving bottom loss values that were 6 dB too low. With this information, Einstein redrew the AMOS bottom loss curves to indicate the best estimate of true bottom loss, showing an amount 6 dB greater than what had originally been shown.[*]

For a directional sonar beam, the surface path near the sonar is excluded, meaning that the bottom loss error involved in the sonar performance calculation was only 3 dB one way but 6 dB in the path

---

[*]With a reflecting surface, there are four combinations of bottom reflection paths that reach a receiving hydrophone, as opposed to only one path in the absence of such a surface. Thus, there is an increase in energy by a factor of 4 due to the surface, assuming that each path contributes the same amount of energy to the received signal. The 6 dB is derived from 10 times the log of 4.

Chapter 3 — Early Concept Formulation

to a potential target and back. The discovery of this somewhat distressing error meant that the original system design calculations were optimistic by 6 dB. If the original analysis were repeated with the revised AMOS curves, the aperture and power would have to be increased by 25%, and pushed the dimensions beyond the realm of the practical.

At this point, Schulkin's bottom loss curves had been overtaken by events. Remembering that Schulkin's curves were the result of a limited statistical sample of what had been observed on a particular group of AMOS measurement stations in the North Atlantic and that these curves provided no indication of location dependence, it can be seen why the AMOS mean values were only an initial estimate and not an accurate representation of the ocean bottom as a whole.

The BRASS system measurement results, on the other hand, permitted a start on the development of an acoustic chart of the bottom that would provide a far more accurate picture of bottom loss statistics. Now a picture of how bottom loss varied with location could be developed, with results that would open up the possibility for selecting routes with better-than-average bottom conditions.

Fortunately, the error in the original bottom loss estimates turned out to have no effect on the original operating frequency selection. The optimum frequency is influenced only by the dependence of attenuation on frequency, which is not affected by the frequency-independent bottom loss error.

NORMAL-INCIDENCE BOTTOM LOSS SURVEY CONCEPT

In December 1957, Nash directed that NUSL prepare a concrete proposal for a program to determine bottom loss as a function of location. After I discussed the issue with Kenneth Mackenzie of NEL, who at the time was the Navy's leading expert on the reflection of sound from the ocean bottom, we both decided that a reasonable approach would be to conduct a bottom reflection survey using a low-frequency echo sounder bouncing sound off the bottom at normal incidence.[34] This information would be used to infer the bottom reflection properties at grazing angles of 10° to 30°, which were those angles of most interest for submarine detection via the bottom reflection path. Upon further investigation, however, it became apparent that bottom loss below a 30° grazing angle was insufficiently correlated with that at normal incidence to make such a survey worthwhile. The British actually undertook a rather extensive survey of the North Atlantic with normal incidence loss measurements but it seemed that little useful information had been acquired.

DEMISE OF SCOUT SHIP CONCEPT

During 18-20 December 1956, I visited Washington, DC, to discuss, among other things, the status of the scout ship. At the time, William Hanley (Surface Ship Sonar Code 848A) wanted to obtain revised NUSL specifications that would include the electrically steered, cylindrical array so that BuShips would be prepared when the funding became available.

At CNO, Captain Sidney Merrill informed me that there was some confusion in CNO about the various proposals from the Navy laboratories. He thought that the NUSL idea for the appointment of an interlaboratory committee to work out a coordinated program was excellent. He also revealed that the concept of an experimental sonar ship was encountering opposition because CNO personnel thought it would be difficult to acquire the funding from Congress for an *experimental* ship. The chances of success would be much better if the ship were proposed as a *prototype*. This advance information meant that the experimental scout ship concept was on its way out and that the candidate platform for the NUSL sonar would now be a more conventional ASW escort ship designed to carry both a sonar and weapon(s).[35] While Commander O'Brien's concept of a "sonar-only" scout ship ultimately met an early demise, it played a key role in creating a constructive attitude on accommodating future ASW ship designs to the NUSL-proposed large sonar array.

IMPLEMENTATION DETAILS FOR THE CONCEPTUAL DESIGN

Search Coverage Method

In February 1957, NUSL's Harold Morrison reported on the outcome of studies concerning the implementation of transmitting and receiving beams for bottom bounce searching with a cylindrical array.[36] The analog techniques that were being used for the beamforming resulted in a large volume and weight. The major reason for such bulk and weight was the requirement for vertical steering of the beams combined with the decision to perform the transmitting beamforming at high power levels. In later designs, each element was provided with its own transmitter, which permitted beamforming at low levels prior to activation of the transmitter drivers.

Only a 30° stepped transmission sector in azimuth per ping cycle was specified for search via the bottom-reflected path. To understand why such a limited sector was chosen, some knowledge of the background regarding the coverage specification is necessary.

The stepped coverage system proposal developed from a growing awareness in the 1950's of the inefficiencies in the HUSL scanning system used in post-war sonars. The scanning sonar developed by HUSL rotated a beam rapidly enough to catch a piece of the echo and present it on a plan position indicator (PPI) display, no matter what the bearing. The beamwidth for this design was approximately 7°. With 360° to cover during the length of the echo, only about 2% of the echo energy was received, which tended to cause a loss of 10 log 0.02, or about 17 dB, for matched-filter processing (where the effective receiver bandwidth is equal to the reciprocal of the pulse length).*

A second problem associated with the scanning sonar design was that the pulsed continuous wave (CW) waveform shape with energy processing was found to perform poorly for a near-zero Doppler target in a reverberation background.[37]

---

*In later years, it was recognized that multiple beams should be formed so that the full echo would be received. Scanning would take place at the output of the receivers for each beam.

A third issue concerned the inability to keep the echo within the dynamic range of the display and receiver. The shorter ranges where reverberation was high were usually overloaded to the point that an echo, even if presented at a high signal-to-background ratio, could not be seen above the reverberation.

A final problem with the HUSL design was that there was no "persistence" on the display between ping cycles, meaning that the echo would often be missed unless the operator happened to be looking at the bearing of the target. Furthermore, the operator could not compare the latest echo reception with those from one or more previous pings.

As a backup to the PPI display, a single audio beam, which was used to confirm an apparent echo seen on the display, was stepped through the forward search sector in a manner similar to that employed on the search-light sonars of the past. A cursor on the PPI display would indicate the direction of the audio beam. While the omnidirectional PPI display was initially thought to be much more effective in search than the narrow-beam stepped audio, the superiority of the PPI was not evident in NUSL's quantitative testing of the operator's search capability. First of all, the operator often followed the audio beam cursor with his eye to be able to hear what he was seeing at each bearing. This approach tended to limit the angular search capability of the PPI to that of the narrow-beam stepped audio. In the second place, the audio channel did not share the display disadvantages enumerated above regarding poor dynamic range and loss of echo energy due to rapid scanning. For these reasons, testing would surprisingly reveal that the audio search performance was, on the average, superior to the PPI performance.[38]

This result initially led me to propose that the audio performance could be further improved by using a four-beam 30° battery of audio beams, instead of only one beam,[39] which would permit increasing the search rate of the audio system. Experiments had indicated that four beams could be effectively covered aurally by using five frequencies (two for the right ear, two for the left ear, and the fifth for both ears). The operator could distinguish bearing to the target and further increase his 5° per ping search steps with a single beam by a factor of 6 for the 30° four-beam sector. When this seemed to work out experimentally, I proposed that a further enhancement would be to use rapidly stepped directional transmission beams to increase the source level in the direction of a potential target. If the aural coverage was going to be the best, why not concentrate the energy over the aural sector and obtain even further improvement?[40]

With all this preceding experience, it seemed only logical for NUSL to use the same technique for covering the bottom bounce path, where the high source level possible with a directional beam seemed to be especially important. However, when the technique was tried for this path, the 30° stepped search did not work well. The long keying rates involved in covering ranges out to as much as 40 kiloyards, coupled with annular range coverage that was sometimes quite narrow, made the stepped coverage unsatisfactory for intercepting a target coming across the coverage annulus. In addition, when a target-like return was received from a piece of the bottom or from a school of fish, the operator would tend to dwell on that bearing for several pings and further seriously slow down the coverage rate. The long range also complicated the problem of correlating a given audio return with one on a previous ping cycle. Later, a sector visual display was provided with improved processing, but only a single-ping cycle history was available per step.

The solution to search coverage (later adopted for the SQS-26 system) was to use a 40° transmission beam with decreased source level. The transmission beam would be programmed to cover a 120° search sector in three rapid steps. In receiving, a battery of 12 receiving beams would cover the 120° sector, with the output of each beam feeding a storage display with a stored history of 6 ping cycles. (The storage would be possible only when improved video processing and storage displays became available.) While this scheme incurred a loss in source level, the greater angular coverage per ping cycle, along with the six stored display histories, would provide significantly improved performance.

Chapter 3 — Early Concept Formulation

Influence of NEL Experimentation on Signal Processing Design

In June 1957, NUSL's Russell Baline visited NEL to discuss the status of its LORAD system.[41] NEL was planning to propose an operational surface ship LORAD that would have a first convergence zone capability for an escorting role and a third convergence zone capability for patrolling slowly on station.* Their proposed sonar platform seemed to be consistent with the scout ship concept, but unfortunately the timing was poor. The scout ship idea was no longer in accord with the thinking in Washington — a problem that arose in part from NEL's location on the West Coast.

NEL personnel informed Baline that *in their LORAD experimentation reverberation was a significant limitation about half of the time*, which seemed to answer the question of whether or not to consider reverberation seriously. NEL personnel discussed with Baline their reverberation-resistant, coded-pulse, broadband processing techniques adapted from radar experience.[42]

In April 1958, Baline documented for SQS-26 application what turned out to be the first proposal for an operational sonar that would use a *combined* frequency-modulated/continuous wave (FM/CW) sonar transmission to cover both low- and high-Doppler targets in a reverberation environment. For low Doppler targets, a wideband FM waveform would permit optimum pulse-compression processing against both noise and reverberation. For high-Doppler targets, the CW-pulse waveform with comb filtering would be the best choice. The first Doppler dividing line between low and high Doppler waveform domains was established at 6 knots, but later it was found useful to lower the Doppler boundary.

This serious attention to processing against reverberation was the first recognition at NUSL of the low-Doppler reverberation masking problems that could be expected with the originally specified single CW waveform. Instead of the single 1-second all-purpose CW waveform, FM and CW waveforms would now be transmitted one after the other to cover both low and high Doppler targets with separate frequency bands

---

*I recall a conceptual sketch of the host ship with a draft of 45 feet.

and receiving systems. This processing system design was strongly influenced by the information Baline had received from NEL, mainly from Dr. James Stewart during his June 1957 visit. Others at NUSL also made important contributions to the implementation of the dual-waveform technique.[43]

Later on, a noise pulse capability was provided as an alternative to the FM waveform. However, while the noise waveform had certain theoretical advantages, it never worked out as well as the FM waveform in practice. Both of these broadband waveforms were called coded-pulse (or simply CP) waveforms. Even after the noise pulse was dropped, the FM pulse was commonly referred to as the CP waveform.

In retrospect, NUSL's initial single 1-second CW waveform design of 1955 seems naïve, but (other than in the pioneering NEL experimentation documented only in the aforementioned 1954 internal NEL memorandum), there had been no attempt elsewhere to use pulsed FM waveforms or Doppler filtering for submarine search in surface ship or submarine sonar. The NRL post-war research in long-range sonar and signal processing techniques summarized in 1955 provided no discussion of wideband FM or noise waveforms other than a brief mention that a noise waveform had been tried. The 1959 paper by Stewart and Westerfield[42] was the first recognition in the open literature of how sonar waveforms should be designed to optimize performance against both reverberation and noise over a range of target Dopplers.

The conclusions from the BRASS experimentation indicated that reverberation did not appear to be a serious problem, which seemed to contradict the NEL LORAD experience that reverberation was a major concern. Variation in operating locations, detection ranges, signal processing, and frequency between the LORAD and BRASS experimental sonars could potentially contribute to the differing conclusions on reverberation masking. It was felt that the primary difference between the two sonars was most likely due to the long coded pulses of the LORAD system, which provided a matched-filtering processing gain against noise of at least 20 dB in excess of that achieved by the BRASS sonar with its short CW waveforms. For this reason alone, the BRASS was much more susceptible to being noise limited (instead of reverberation limited) than was the LORAD system. Of course, BRASS was still able to detect bottom bounce echoes despite its poor processing against noise, but only

Chapter 3 — Early Concept Formulation

against a beam-aspect target submarine, where the target strength was on the order of 10 to 15 dB greater than it would be for nonbeam aspects.

PROCUREMENT PLANS

Contracting for Experimental Equipments

NUSL forwarded the revised specification for the cylindrical array to BuShips on 4 January 1957.[44] On 10-11 January, Harold Nash and the NUSL Commanding Officer, Captain Harold E. Ruble, made a follow-up visit to BuShips and CNO, where they met with Commander William Hudson (Code 845), William Hanley (Code 848A), and Lieutenant Bradford Becken (Code 848A).

NUSL was hoping to have funds allocated for at least preliminary study work on the sonar in anticipation of FY58 formal funding. A consensus was reached that Code 848A would prepare specifications for a procurement by the end of February, using the NUSL specifications as input. Partial FY57 funding would be sought to begin the procurement. One-on-one discussions with potential bidders would in any event be initiated.[45]

Nash and Ruble brought up the subject of a BuShips, NUSL, NEL, and NRL committee, resulting in a letter being drafted at once to charter the committee. An initial meeting was scheduled for 28-29 January 1957. Leo Treitel (BuShips Code 845) would be chairman; members would be Nash (NUSL), Donald Wilson (NEL), and Harold Saxton (NRL).

At CNO, Nash and Ruble called on Rear Admiral Frederick Warder (OP-31) to discuss the status of NUSL's proposed sonar with a number of the key captains and commanders in his organization.* All seemed to support proceeding with the sonar, but the NUSL contingent was somewhat initially disconcerted to learn that the request for the emergency funds recommended by Project NOBSKA (a summer study of undersea warfare at Woods Hole) had been turned down by the CNO Research &

---

*Warder was dubbed "Fearless Freddy" for his aggressive tactics in the Pacific during World War II. As skipper of *Seawolf* from 1941 through 1942, he was one of that small band of submariners who vainly attempted to stem the tide of Japanese conquest. Despite the existing torpedo design problems that were not corrected until late 1943, Warder sunk eight Japanese ships.

Chapter 3 — Early Concept Formulation

Development Review Board, which recommended that BuShips fund the sonar out of its own budget. Fortunately, this decision had no adverse effect on BuShips' progress toward procurement of a system, and, in fact, the NOBSKA recommendation that the proposed system should be funded actually seemed to have had a favorable influence on the eventual procurement decision.

In the spring of 1957, a bidder's conference was held at NUSL for the sonar based on the January 1957 NUSL specifications for the cylindrical array system.[46] In September, BuShips issued formal specifications for a sonar system to be known as the SQS-26, and, in November, the bidders' proposals were received. BuShips decided to procure two experimental systems: a conservatively designed XN-1 model from EDO Corporation and a more innovative XN-2 model from GE. NUSL recommended that once the contracts were let, EDO should concentrate on the early delivery of a system with a bottom bounce capability; GE would be allowed more time to incorporate both a bottom bounce and a convergence zone capability.[47] This approach was formally recommended to BuShips on 23 May 1958.[48]

A feeling for the fluid nature of the specifications is indicated in the documentation of the 23 May 1958 meeting at BuShips with NUSL, BuShips,* and GE.[49] Because GE had not received Baline's FM/CW waveform proposal at this point, Baline proceeded to describe it. Elmer Landers, who was to be the BuShips civilian manager of the SQS-26 development, then mentioned that the SQS-26 specifications were still not complete — this was only a month before the contract award. GE was informed at this time that it would be expected to provide a fully integrated convergence zone and bottom bounce system.

In June 1958, contracts were finally let for two AN/SQS-26 systems. Since the combined FM-CW waveforms in Baline's April 1958 proposal were not included in the original specifications, the contracts were modified soon after signing to include this feature, which was now believed necessary to combat reverberation.[50]

---

*Lieutenant Commander Bradford Becken was now the military head of BuShips Sonar Code 848A (and also had the distinction of having been in previous years part of the Rhode Island Boy Scout troop of which Bill Downes was the scoutmaster).

Chapter 3 — Early Concept Formulation

An early meeting with EDO was held on 9 July 1958, shortly after the contracts had been awarded. At this time, the originally proposed 1-second CW waveform was still being discussed. The system was continuing to be designed after the contract award, in respect to both the original specifications and implementation plan.[51]

In the late summer of 1958, GE modified the SQS-4 on USS *Murray* (DE-576) to transmit pulses and receive submarine echoes using both FM and pseudorandom noise (PRN) waveforms.[52] The intent was to verify that these coded pulse waveforms could be successfully received through an ocean medium and processed with a delay line time compression (DELTIC) correlator.[53] The waveforms and processing were similar to what had already been experimented with in the NEL LORAD system.

On 5 September 1958, a 3-day post-contract award meeting was held at NEL, with NUSL, GE, and EDO represented. Dr. James Stewart, Frank Hale, Melvin Pederson, and James Whitney of NEL and Victor Anderson of the Marine Physical Laboratory, Scripps, presented their LORAD experience. Discussed were displays, signal processing, waveforms, reverberation, transducers, and shallow-water results.[54] Hale and Pederson mentioned that the reverberation encountered is highly variable, but correlates with season. It is low in summer and fall, but high in winter and spring. Later, it would be concluded that this seasonal variability was produced by marine life. However, it was not until the early 1960's that the effects of marine life on sonar performance were beginning to be fully understood.

In early 1959, problems began to surface on the course of the SQS-26 development. For one example, NUSL's Downes reported on extensive discussions of displays with BuShips during which time it was agreed "that the GE paper tape recorder is unacceptable."[55] Because there was no "quick fix," the paper tape display (used for bottom bounce and convergence zone path coverage) was delivered with the initial XN-2 configuration on USS *Wilkinson* (DL-5). It was also included in the first buys of the AX production systems.

By July 1960, a number of changes to the two development contracts were in various stages of implementation. In addition to the aforementioned FM/CW waveform and processing changes, modifications were made to (1) incorporate smooth transmission stabilization in depression angle, (2) provide compatibility with the antisubmarine rocket (ASROC) fire control system, (3) incorporate passive detection, and (4) add a sector scan indicator (SSI). In addition, BuShips was in the process of contracting for (1) preformed video beams for surface duct path omnidirectional coverage, (2) individual transistorized power amplifiers for each transducer element, and (3) the study and development of new displays.

Production Model Decisions

Once the two-system contract was let for what were experimental versions of the SQS-26, Washington planners started to lay *production* plans that would require more SQS-26 procurements before the first two systems had been to sea. In February 1959, CNO's long-range objectives group (OP-93) had asked BuShips to investigate the possibility of building inexpensive escorts with the SQS-26 sonar and a companion weapon. Inexpensive escorts were defined to be the "minimum platforms capable of supporting an SQS-26 sonar and a means (generally a helicopter) for prosecuting its contacts."[56]

In March 1959, BuShips was already planning to contract for several production systems.[57] Both EDO and GE were directed to prepare specifications for a potential production contract award in October 1959. The candidate platforms would include two FY60 "ASW ocean escorts" (*Bronstein* (DE-1037) and *McCloy* (DE-1038)), as well as the experimental escort *Glover* (AGDE-1). The *Glover* was later postponed until the FY61 building program.[58] The two escorts would be the only production systems to be designated as "SQS-26"; all future production systems were to append AX, BX, or CX to the SQS-26 designation.

According to NUSL's Downes, the Navy made a "momentous decision" for the SQS-26 program in August 1959:

> The Ship Characteristics Board, faced with deciding what sonar would best support the ASROC weapon system planned for new destroyers, chose the SQS-26 to be a good match, as compared to the SQS-23. The Board recognized that unproved first models of the SQS-26 would have problems that would require backfitting hardware changes. Nevertheless, they decided that it should cost less to install the SQS-26 and backfit changes than to install the SQS-23 and replace them later on.[59]

Chapter 3 — Early Concept Formulation

This decision would open the floodgates to a deluge of SQS-26 production systems in future years. The Department of Defense policy at that time required large numbers of ASW surface ships for the potential defense of convoys to Europe.* The sparest projections called for 12 convoys, each requiring 10 to 12 escorts. Thus, although NATO navies required a minimum of 120 escorts, counts of available NATO ships found fewer than half that number available. The FY62-FY67 budgets funded 60 surface warships designed principally for ASW, which was a fivefold increase over the previous decade.[60] Each of these ships was to be fitted with an SQS-26 sonar.

IMPORTANCE OF SYSTEMS ENGINEERING FUNCTION TO SQS-26

In August 1959, I was reassigned to Harold Nash's staff to work for Stanley Peterson, who was forming a Systems Planning Staff under Nash. At the time, Nash was head of the Sonar Systems Development Department, which included surface, submarine, and fixed sonar projects.

In this new assignment, I was responsible for formally providing a "systems engineering" function for all the sonar projects in the Sonar Systems Development Department. For the immediate future, I would continue to support the SQS-26 program, which was then at a particularly critical stage in its development.

Although the term "systems engineering" was not commonly used in the research and development community in those days, it best describes what was being done at the NUSL staff level. Since then, systems engineering curricula leading to masters degrees have been introduced in many engineering schools, although it was not until after 1997 that a commonly accepted definition of systems engineering began to appear in the literature.[61]

The type of systems engineering carried out at NUSL in 1959 is best defined in a 1965 paper by Dr. James B. Fisk, who at the time was

---

*Convoys would be needed to supply the North Atlantic Treaty Organization (NATO) if the Soviet army were to initiate a conventional war in Europe. In 1942, Britain had brought in 35 million tons of cargo from abroad; in 1965, the NATO countries in northwestern Europe alone imported 400 million tons.

President of Bell Laboratories, where the technique seems to have originated several decades earlier. Their method happened to fit almost exactly what NUSL was referring to at the time as "systems planning" and later as "systems analysis." At Bell, the systems engineering effort was carried out independently, under its own vice president. Key excerpts from Fisk's description of the various functions carried on in the systems engineering organization are as follows:

> In the systems planning function, appraisals are made of the various technical paths that can be followed to employ the new knowledge obtained by research in the development and design of new systems . . . . As the technology of communications and of weaponry is broadened and becomes more complex, the choice of the technical paths to be pursued in the utilization of the new technology has become increasingly difficult. It is this situation that has led to the evolution of systems engineering as a means of guidance.
>
> Systems Engineering also maintains close association with the work of our research organization and knows intimately the content of our new knowledge reservoir. It integrates the knowledge from operating experience, research, and development, and with this as a background, makes sophisticated studies that appraise development projects for new systems and the apparatus required for these. Each study defines the objectives of the proposed development, describes the functional performance of equipment or systems that are needed, and often describes a particular embodiment of the system by way of example.
>
> As the development organization proceeds with a project, systems engineers maintain close contact, continuously observe the technical difficulties encountered . . . amend the objectives and plans as required . . . organize field trials often needed during the course of development, and are responsible for the tests and the evaluation of results. When a system is standardized and new equipment placed in manufacture, Systems Engineering, together with the development group, follows service performance of first installations and coordinates the "growing pains" that accompany new systems as they enter service. It finally participates in the evaluation of the service and its economic or military worth.
>
> The staff members in the systems engineering organization . . . are largely men drawn from [research, development, or operations], who have exhibited special talents in analysis and the objectivity so essential to their appraisal responsibility. With respect to the contacts made with the research organization, it should be evident that such contacts made judiciously by and with the right individuals can provide for research

people an atmosphere of encouragement and stimulation without annoying pressure or semblance of dictation.[62]

In 1966, I was appointed head of the Systems Planning Staff, which reported to Peterson who had been promoted to Associate Director for Plans and Programs. By that time, Nash was Technical Director of NUSL.

With the formation of the Naval Underwater Systems Center (NUSC) in 1970, the Systems Planning Staff became the Systems Analysis Department, continuing to perform a systems engineering function for the Center (and eventually growing to 40 personnel).* I continued as head of that group through the completion of the SQS-26 development work in 1975. By that time, Harold Nash had retired, and a new administration decided that the department should be broken up and the functions and personnel distributed among other departments at NUSC.

Although the main responsibility of the Systems Analysis Department was systems engineering as it has been defined above, far more than the systems engineering for the SQS-26 was carried out. A few examples are provided to highlight some of these efforts:

- In response to a request from the first Polaris submarine, the *George Washington*, I wrote a sonar performance prediction manual for its initial deployment. The manual later became popular in the submarine force, with 700 copies printed. For many years, I continued to encounter both active and retired submariners who knew me mainly as the author of that manual.

- Herbert Fridge spearheaded conceptual design studies for the BQQ-5, which became the standard sonar for SSN 688 class submarines.

---

*Personnel in my department who made key contributions to the SQS-26 development included Bernard Cole, John Hanrahan, Eugene Podeszwa, Harold "Joe" Doebler, Richard Chapman, Juergen Keil, George Brown, Gustave Leibiger, Herbert Fridge, Carlton Walker, and David Williams. Others working in the Sonar Development Systems Department contributing first-class "systems engineering" functions to the SQS-26 program were William Downes, Russell Baline, John Snow, Frank White, and Walter Hay.

Chapter 3 — Early Concept Formulation

- In 1968, the department brought aboard Michael Pastore, a former submarine officer, to further strengthen NUSL's submarine systems engineering capabilities. He pioneered the use of narrowband analysis techniques for the hull-mounted submarine arrays and later applied his passive sonar expertise to the surface towed array sonar system (SURTASS) arrays being introduced in surface ships. With NUSL's Jan Holland, Pastore organized and participated in experimental studies at sea of low-frequency noise sources, identifying "hot" spots related to shipping patterns and developing search tactics to avoid interference from these noise concentrations.

- William Wardle proposed the first techniques for installing prefabricated arrays of hydrophones embedded in rubber blankets onto submarine or surface ship hulls. He also carried out many innovative studies of low-frequency, active surveillance system options.

- Gustave Leibiger specialized in the development of propagation loss models for use in both submarines and surface ships. His RAYMODE (a mixture of ray and normal mode theory) was adopted as the official Navy model for use in shipboard performance predictions. Leibiger also designed an integrated passive performance prediction system for Trident submarines.

ANALYSIS OF SQS-26 SYSTEMS FOR OPEN-OCEAN SEARCH

In 1959, Stanley Peterson and I had been asked by Albert Bottoms, who was then with the Weapons Systems Evaluation Group (WSEG) on the staff of the Secretary of Defense, to take part in a WSEG study of defense against Soviet ballistic missiles. This problem was typical of what WSEG was set up to analyze, cutting across Navy, Air Force, and Army components of the Department of Defense. NUSL was to look at the part of the problem that was concerned with defense against ballistic submarines off the shore at locations close enough to potential targets in the continental United States to create the special situation of low-warning time between the missile firing and target impact. My contribution was to examine the possibility of forming an ASW sweep capability using a mix of submarines and surface ships, all having long-range sonar capabilities that were expected to be developed within the next few years.[63]

Figure 7 shows the sweep lines from the postulated mix of ASW detection systems that NUSL believed could be made available by 1963.[64] The 800- by 1500-mile rectangle represented a typical area in which missile-launching submarines might be stationed. It was estimated that each sweep line transiting the entire area would have about a 33% probability of detecting a given submarine.

With John Hanrahan's mathematical help, it was determined that this assumption would lead to an average time for the detection of a given submarine of 7.8 days, which, of course, would represent a Cold War scenario. If each submarine detected could be tracked in an average of 2.6 days, it would mean that an average of 25% of the submarines were being tracked by platforms capable of delivering an attack, with the help of air support (25% would be a statistical figure that could fluctuate from 0 to 100%). The objective would be to weaken the confidence of the Soviets in their capability to deliver a successful missile attack. Of course, this was not a prediction of what could be done in this early stage of long-range active sonar development, but only a suggestion of a possible useful application of the new sonar techniques that were being investigated.

Some of the senior people at NUSL were quite critical of the study because it seemingly contradicted the NUSL policy at the time that fixed, active, bottom-mounted sonar provided the best solution to offshore surveillance against quiet Soviet submarines. This criticism was far from unanimous, however. Dr. Horton, in particular, enthusiastically endorsed this new study as an approach that should be considered. Later, Raytheon's distinguished Laurence Batchelder was interested enough to make a special trip to NUSL to discuss the report with me.

The events to be related in chapter 9 would show that a decade later the concept of performing a large-area ASW search with active sonar and follow-up submarine tracking was not as fanciful as it might have appeared in 1960. As a matter of fact, in the early 1970's, the Fleet would repeatedly demonstrate the capability of performing a one- to two-ship ASW area search and hold operation using the convergence zone mode.

Chapter 3 — Early Concept Formulation

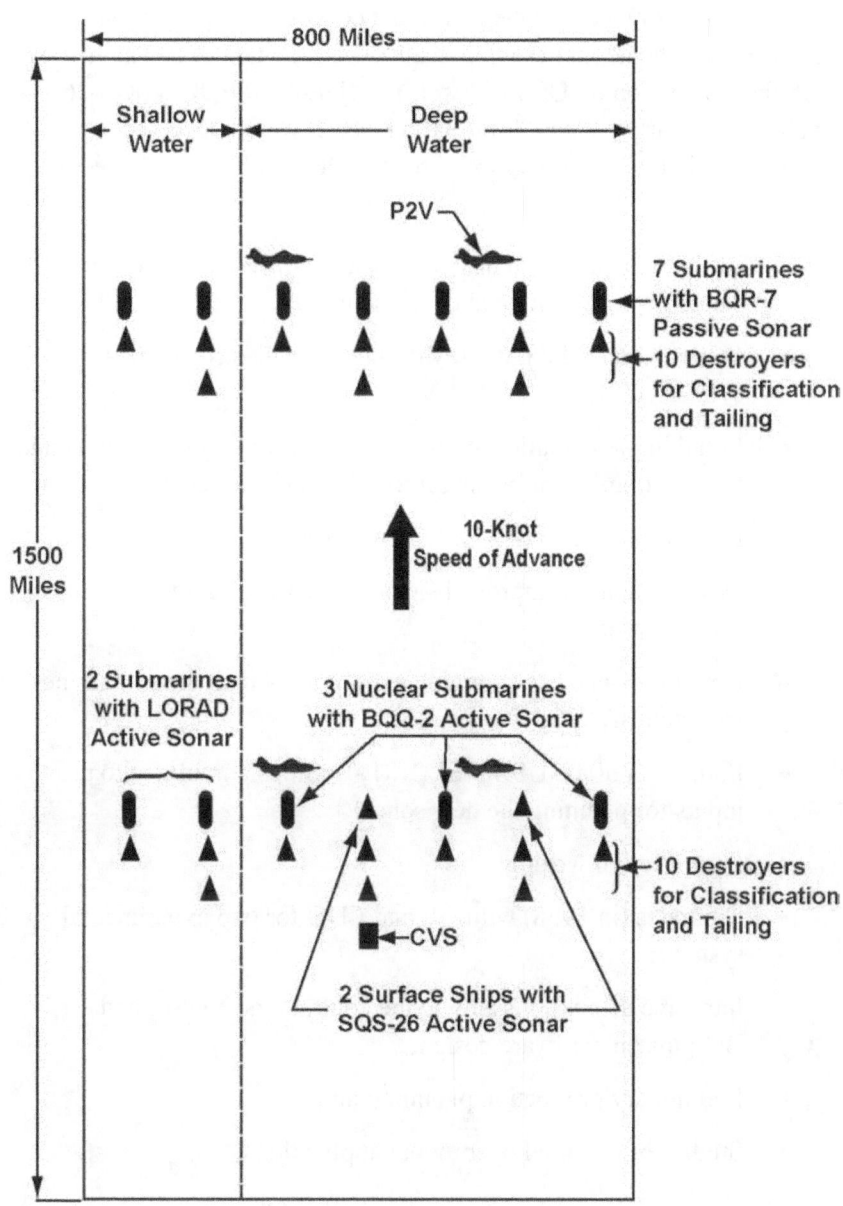

Figure 7. Notional Formation of Long-Range Active Sonar Ships for ASW Search of a Large Ocean Area

Chapter 3 — Early Concept Formulation

DEVELOPMENT STATUS AT THE END OF 1960

The year 1960 marked the end of the concept formulation phase of the SQS-26. The next phase, full-scale experimentation, began in 1961 with the installation on USS *Willis A. Lee* (DL-4) of EDO's SQS-26 (XN-1). The early concept formulation phase, which occupied approximately 6 years from 1955 through 1960, had accomplished the following:

- An initial conceptual design with a flat-face array, mechanically steerable in azimuth and depression;

- Conversion of the flat array design to an electrically phased cylindrical array;

- Initial implementation studies of waveforms, signal processing, transmitting, beamforming, receiving, and switching;

- Discussions with NEL on LORAD results;

- Development of improved signal processing against reverberation;

- Bottom bounce echo-ranging experiments with the submarine-mounted BRASS;

- Formation of a NUSL-NEL-NRL-BuShips committee to provide inputs for planning the new sonar;

- Specification writing;

- Contracts (in 1958) with GE and EDO for two experimental systems;

- Intensive discussions among the contractors, NUSL, and BuShips on hardware design;

- Preliminary production planning; and

- Studies of potential operational applications.

There was no realization of the magnitude of the task that lay ahead in the attempt to build an echo-ranging sonar system to exploit surface duct, bottom reflection, and convergence zone long-range sound paths. First of all, a complex equipment design involving previously untried techniques would be required. Secondly, solving the problems of

Chapter 3 — Early Concept Formulation

optimum operation of the equipment on each of the three long-range paths (about which there was incomplete knowledge) would require a major effort. Furthermore, the only decisions required by the operator of the older sonar systems were the selections of keying scale and pulse length. For the SQS-26, an operator would also be required to determine the following in any given shipboard environment:

- Estimated capability available on each long-range path,
- Optimum depression angle for each path of interest,
- Equipment range window for the zones formed by the bottom bounce or convergence zone paths, and
- Average sound speed for each of the paths to permit a conversion from elapsed time to range.

All the foregoing tasks required a computation that used the following environmental input information (which had to be either measured or estimated):

- Sound speed versus depth all the way to the bottom,
- Water depth,
- Wind speed as it affected sea surface scattering,
- Biological scattering strength,
- Bottom reflectivity, and
- Bottom-scattering strength.

High-speed shipboard computers were not available in those days to make computations of the optimum equipment settings (even if such input information had been available). It was therefore necessary to develop rules of thumb and cumbersome tables of predigested computations for system operating guidelines, both of which were only partially successful, even in the presence of a skilled operator. The difficulties that an operator would have in setting up the equipment to exploit deep sound paths were as imposing in 1960 as they were to Urick 9 years earlier.[65] In fact, with an increased understanding of the difficulties, the problem seemed even more formidable.

CHAPTER 4
FULL-SCALE EXPERIMENTATION AND DEVELOPMENT

TWO SQS-26 EXPERIMENTAL SYSTEMS

A technical evaluation was set up under project T/S-25 in 1961 for the shipboard testing of EDO's SQS-26 (XN-1) on USS *Willis A. Lee* (DL-4). Then, in 1962, another technical evaluation was arranged under project T/S-26 for the testing of GE's SQS-26 (XN-2) on USS *Wilkinson* (DL-5). Both systems were expected to receive an operational evaluation after the technical testing was completed. While each technical evaluation was originally planned only as a test program, equipment deficiencies were uncovered during the testing that required correction. The program therefore assumed a considerable amount of development activity in addition to the technical testing.

Since the development of the first scanning sonar at HUSL during World War II, an evolutionary succession of improved sonar systems had been successfully introduced in the post-war years with little or no development work undertaken at sea. When the SQS-26 was introduced, it was hoped that this system would be treated in the same fashion since, in many ways, it was also an evolutionary step. In reality, the attempt for the first time to exploit sound paths that extended throughout the entire depth of the deep ocean involved techniques and phenomena that were so new that much more development and experimentation at sea would be required.

Need for Full-Scale Experimentation

The reader may wonder why the SQS-26 program required a "full-scale experimentation" phase when its performance had already been demonstrated with full-scale experimental sonar equipment (the BRASS system described in the last chapter). The BRASS experimental system indeed gave timely and convincing proof that bottom bounce echo-ranging performance independent of thermal conditions out to ranges of about 20 kiloyards in deep water was possible. Given a reasonably reflective bottom, consistent echo returns could be shown, *at least on a beam aspect target*. Furthermore, a beginning was made on the selection of favorable locations that could serve as the basis for planning convoy

Chapter 4 — Full-Scale Experimentation and Development

routes upon which reliable long-range detection performance would be expected.

The BRASS experiments had strongly influenced Navy decisions not only to go ahead with the procurement of two experimental surface ship sonars, but also to go into limited production to provide sonars for the future construction of ASW escorts. These experiments also pioneered the development of important experimental techniques for instrumenting target submarines to permit ping-by-ping measurement of propagation loss during echo-ranging tests.

Despite the value of the BRASS experimentation, there were many limitations — some were obvious and others were not. Many of the important issues that had not been explored in the BRASS program became more apparent with the passage of time. Summed up, it was felt that BRASS did *not* demonstrate the following:

- Performance against a nonbeam aspect submarine.

- Successful search for a target at an unknown bearing and range.

- The effectiveness of matched-filter processing with a coded pulse.

- The effect of the environment on reverberation with the use of coded pulse processing.

- The effectiveness of filtering out low Doppler CW-pulse reverberation with high Doppler CW-pulse targets.

- Beamforming and beam steering for a fixed multi-element array of the type that would be required aboard a surface ship. (Of concern were the effects of element interaction on the phase and amplitude required for each element to form a beam at any steered angle of interest.)

- Beam-switching techniques for both transmitting and receiving beams.

- The mechanics of displaying target information as a function of vertical search angle and range.

- Carrying out of a convergence zone search.
- A dome design that could maintain satisfactory flow noise levels at operational speeds.

All the above problems had to be explored with *both* cylindrical array SQS-26 experimental systems — XN-1 and XN-2. While the limitations in the BRASS experiments are clear in hindsight, they were by no means generally agreed upon during the BRASS program. Some of the major issues at that time are summarized in the following paragraphs.

Because a *beam aspect submarine* had been used for all tests, the effect of the vertical arrival angle on target strength was unknown. Some NUSL personnel held the optimistic view that the vertical arrival angles would give target strengths at all aspects comparable to those experienced at beam aspect with horizontal arrival angles. If this belief were accurate, it would increase the value and the generality of the beam-aspect testing results. Later testing revealed that the target strength significantly decreased as the azimuthal angle of incidence departed from the beam aspect geometry, even for vertical angles of 30°, which were used in the BRASS experiments.

The use of *coded pulses,* still novel in the 1950's, was another controversial topic. While NEL's LORAD experiments employed coded pulses and matched-filter processing in the early 1950's under the leadership of Dr. James Stewart, there was a widely held view at NUSL that degradations from medium effects would make the processing of coded pulses impractical. Coded pulse processing originated in the radar field, but, even there, key experiments on FM waveforms with pulse-compression matched-filter receivers did not begin until 1951.[1] While coded pulses were incorporated in the SQS-26 specifications in 1958, with the help of information received from NEL, it was not until the mid-1960's that NUSL began to acquire significant expertise in the performance of pulse-compression processing. The use of short pulses on the BRASS system obtained reasonable results for beam aspect targets, but the omission of experiments with long coded pulses and matched-filter processing resulted in noise masking for the BRASS system that was serious enough to obscure reverberation and prohibit the quantitative investigation of reverberation levels as a function of environment.

Initial Tests on the SQS-26 (XN-1)

EDO's SQS-26 (XN-1) was installed on *Lee* in early 1961. After "debugging" the equipment and making basic calibration measurements, submarine echo-ranging experimentation began in mid-1962. It would be exciting to finally see how successful NUSL's 1955 concept of performance would be after some 7 years of trying to convince the Navy of its viability.

From 20 to 26 June 1962, NUSL conducted controlled testing of bottom bounce echo-ranging on a beam aspect submarine target at ranges between 20 and 30 kiloyards. The general methodology of the tests was designed not only to determine echo-ranging effectiveness, but also to measure reverberation, noise, minimum detectable signal, and propagation loss so as to ensure an understanding of all elements of the sonar equation — an approach that was adhered to in all future SQS-26 testing. The test area (located between Long Island, New York, and Bermuda in 2700 fathoms of water) had already been investigated during BRASS experimentation. The NUSL results for bottom reflectivity measurements and bottom bounce echo-ranging effectiveness beyond 20 kiloyards were found to be consistent with the BRASS experiments. However, thermal conditions were unfavorable for conventional *direct path* echo-ranging at more than a few thousand yards.

Bottom bounce detection ranging out to 30 kiloyards from an operational surface ship was an echo-ranging "first." Lieutenant Commander Richard Duggan, an on-site representative of the Key West Development Detachment of the Navy's Operational Test and Development Force, was significantly impressed with this long-range echo-ranging performance under such unfavorable thermal conditions. At times, the thermal structure presented negative gradients of 10° per 100 feet, conditions under which no surface duct path could possibly exist.

In other ways, however, the results were disappointing. With FM processing, the dominant interference was reverberation, the origin of which was not obvious at the time. Such interference reduced the performance over that which would be obtained on the system if noise had been the limiting background. Furthermore, if a search over a wide azimuth sector were attempted, other equipment losses in azimuthal beam coverage and display of echoes would be incurred. Nonbeam aspect submarine targets would clearly be a problem.

Finally, equipment reliability was poor. Downtime while repairs were being made was a continual interruption to the testing. Even when the system was operating, it was difficult to know how well it was working for any particular condition. Interminable checking of the equipment was required to be certain that the sonar measurements were valid. With such a large and complex system (576 transducer elements and associated beam steering in two dimensions), it was a struggle just to monitor equipment performance. It became evident that there were many unresolved problems with the electronics that would require extensive investigation.

Initial Tests on the SQS-26 (XN-2)

In November 1962, submarine echo-ranging tests with the SQS-26 (XN-2) on *Wilkinson* began in deep water off the coast of California. The test area was 2200 fathoms deep off Point Conception, south of Monterey, which was the same location that had been used by NEL for LORAD testing. As in the XN-1 tests, measurements were taken for reverberation, noise, minimum detectable signal, and propagation loss, as well as for echo-ranging effectiveness so as to ensure a measurement of all elements of the sonar equation. Some testing was also conducted on transponder signals transmitted from a surface ship to simulate submarine echoes.

The NUSL demonstration of convergence zone echo-ranging was a first from a U.S. Navy operational ship. Seven runs against a submarine target were made for convergence zone detection at beam aspect and one at a 45° aspect, with all eight runs successful. While time did not permit bottom bounce testing, expectations were good based on minimum detectable level and propagation loss measurement results.

Further testing planned for December and February had to be aborted because of equipment casualties involving transmitter, switching, transducer, and programming failures. As found in the XN-1 testing, equipment reliability was disappointing.

Chapter 4 — Full-Scale Experimentation and Development

In March 1963, further XN-2 echo-ranging tests off California were carried out over surface duct, bottom bounce, and convergence zone paths. The surface duct detections were poor because of adverse thermal conditions. The bottom bounce and convergence zone detections were not as strong as expected from the November measurements in the same location because the reverberation was 10 to 15 dB higher in March than it was in the previous November.

The major increase in the reverberation from November to March was initially as puzzling as it was dismaying. However, discussions at NEL following the testing provided valuable information on the likely explanation for such problems. Ken Mackenzie had earlier found that LORAD reverberation off California, which showed a seasonal dependence, corresponded with an annual cycle in phytoplankton production. Peak reverberation on LORAD was observed in March, the time of year that NUSL had unfortunately selected for its important XN-2 testing in the same general location.

A discussion with Dr. Eric Barham, a NEL marine biology expert, identified the likely source of the problem. Barham had studied deep scattering layers in the same location in the previous year using the bathyscaph *Trieste* for both acoustic and visual observations. He had visually identified a mysterious acoustic scattering layer evident on an echo sounder as an aggregation of Pacific hake larvae with air bladders large enough to produce a high level of backscattering.

A 1955 Fish and Wildlife Service paper by fishery experts Ahlstrom and Counts indicated that the population of the Pacific hake larvae tended to reach a maximum in March in the NUSL test location.* Moreover, this particular location historically tended to have a higher concentration of larvae than did surrounding areas.[2] Typical numbers for hake larvae per standard haul in the years 1951-52 were 3 in January, 21 in February, > *200 in March,* 20 in April, and none in May. The Ahlstrom and Counts data fit well with the information from Mackenzie

---

*The Fish and Wildlife Service had extensively studied the biological population in the waters off California as a function of season and location. These studies were motivated by the disappearance of the sardine species that in the 1930's had provided a thriving industry in the Monterey area. (Steinbeck had given prominence to that industry in his classic novel *Cannery Row*.)

and Barham, and together they provided a plausible explanation for the 10- to 15-dB higher reverberation in March than in November.*

The test schedule allowed only limited investigation of the bottom bounce path. Moreover, the reverberation from the aforementioned hake larvae was troublesome for both the bottom bounce and convergence zone paths. Although the layer of hake could be resolved by a vertically oriented echo sounder, the geometry of long-range submarine detection was such that the sonar beam simultaneously included both the submarine echo and the competing marine life reverberation.

In some locations, propagation loss via the bottom was too high to permit the successful detection of echoes, but in other locations it was quite favorable. Good agreement was obtained between results with injected artificial echoes and real submarine echoes, suggesting that any distortion produced by the bottom bounce path did not measurably degrade the coded-pulse signal processing.

Unreliable equipment was still a problem, as it had been in previous XN-2 testing. The equipment characteristics were uncertain from hour to hour, and the equipment sometimes had to be shut down completely for repairs. At other times, hidden degradations would occur during the testing. The struggle to know the equipment status at any particular time resulted in basic equipment checks prior to, during, and after submarine echo-ranging.

Criticism of Measurements with the Two Experimental Models

One of the questions that had been raised by critics of the program was the following: How could anything be learned from experimental tools that were as complex and unreliable as were the two SQS-26 systems? The proponents of this argument felt that simpler systems should be used to obtain basic information that could then be used to accurately infer the performance possibilities and limitations of more complex equipment *before* it was built. The problem with this alternative is that predictions of what will happen with complex

---

*As scientifically fascinating as this information was, the negative impact on the echo-ranging experimentation seemed to be the classic fulfillment of Murphy's Law — if anything can possibly go wrong, it will.

equipment cannot be accurately made without understanding the characteristics of that equipment. *Obtaining this understanding is possible only by observations on the full-scale equipment, with all of its unpredictable interactions.*

That is not to say that other measurements on simpler systems are not of value. However *total system observations* are essential in determining the validity of models that relate equipment performance to all other pertinent sources of knowledge. By thorough equipment monitoring and the use of experimental sampling techniques that permit estimates of statistical significance, NUSL was able to obtain meaningful measurements. It is true that much effort was exerted to ensure that the results were not affected by equipment problems. On the other hand, this type of vigilance is necessary in any experimental investigation.

MANAGEMENT CONCERN WITH EQUIPMENT RELIABILITY AND PERFORMANCE

While the scientific results were yielding significant information, by the end of 1962 the unreliability of the XN-1 and XN-2 systems was alarming Navy management. Concern was heightened by the Navy's previous decision to proceed with production prior to completing tests on the two experimental systems. Production contracts had been placed in 1960 for 2 SQS-26's, in 1961 for 12 SQS-26's (AX), and in 1962 for 18 SQS-26's (BX). The rationale for the early release to production was based on the expectation that these systems would have capabilities superior to those of the SQS-23. While there was a recognized risk that performance would not work out as anticipated, changes could always be retrofitted. In the meantime, it was felt that the systems would at least perform as well as the SQS-23. However, with the excessive casualties that were occurring in both experimental units so far, it seemed that the same reliability problems might surface in all 32 production systems.

Establishment of the SOFIX Program

Faced with the potentially disastrous problem of having unreliable sonar systems on the new ASW ships, Navy management, in March 1963, took a number of steps to bring the situation under control:

Chapter 4 — Full-Scale Experimentation and Development

- A special project office was established in the Naval Ship Systems Command to (1) address the equipment reliability and performance problems now evident in the two experimental systems and (2) handle the necessary changes to existing production system specifications. The office, called SOFIX for "Sonar Fix," would be headed by Captain William Peale and his civilian counterpart, Elmer Landers. The office later became PMS 387.

- A "development assist" project, D/S-281, was set up to develop and test necessary modifications to the XN-1 and XN-2 systems. This was the first formal recognition that more development work was required, as opposed only to testing what had already been developed. The XN-2 would become the prototype system.

- NUSL was provided with more funding and the authorization to hire the necessary manpower to properly carry out the increased effort. From 1955 through 1959, the manpower at NUSL on the SQS-26 program had not exceeded 2 man-years per year.* With the establishment of SOFIX in 1963, it grew to an 18-man-year rate as a result of Washington funding and billet allowance increases. Over the next few years, manpower continued to increase until it peaked in 1966 at 89 man-years per year.[3]

- Contracts were established with GE and Tracor for assistance in evaluating the existing signal processing design and potential modifications.

- An intensive review effort of the production system designs was started at NUSL, with contract assistance from RCA. The effort peaked at 26 NUSC and 8 contractor engineers.

---

*The low allocation of manpower to NUSL for the SQS-26 project during its first 5 years was the consequence of a decision made prior to 1955 by NUSL management. Research and development efforts at NUSL were to be concentrated on fixed, bottom-mounted active sonar, which would require a nearly 100% manpower commitment both in the surface ship sonar development and sonar research departments. The name of the Downes organization was changed to the Surface Ship and Surveillance Sonar Department to reflect the change in department responsibilities. Through 1959, Russell Baline had been the only one working full time at NUSL on the SQS-26 program.

- A pressurized rubber dome development program was initiated to overcome the problems observed with high acoustic transmission levels that were causing steel dome paint erosion and consequent noise problems. B. F. Goodrich — which had built rubber domes for the smaller sonars during World War II — first proposed the rubber dome for the SQS-26 sonar array in February 1963.

XN-2 Refurbishing

Under the direction of the SOFIX office, a major effort was undertaken at GE (led by Ken Greenhalgh and Kyrill Korolenko) to correct deficiencies in the XN-2 equipment on the *Wilkinson*. From 30 March to 15 June 1963, 10 to 15 GE engineers and technicians worked on the ship at the Long Beach Naval Shipyard in an effort to provide a more reliable, better performing XN-2 system. After this work was completed, the ship moved its operation base from the West Coast to the East Coast, with all further testing on *Wilkinson* performed in East Coast test locations.

NUSL CONCERN ABOUT MAINTAINING AND OPERATING PRODUCTION SYSTEMS

Looking beyond the experimental systems to the production units, NUSL's Downes was becoming concerned about the capabilities of shipboard personnel *to both maintain and use the systems,* even after the reliability issues were overcome. Problems in maintaining and using the two experimental systems, even with experienced engineering personnel, already existed. On 2 April 1963, he expressed his uneasiness to SOFIX management in Washington.

There was hesitancy at the SOFIX office to address the training of personnel. They felt that "it would probably take the development of a serious situation before various people would be willing to change the existing setups to the extent necessary."[4] The fundamental problem (in this author's view) seemed to be that *training* was not something that was normally part of the responsibility and interests of either NUSL or BuShips. A separate Bureau of Personnel (BuPers) was responsible for sonar training in conjunction with the East and West Coast Fleet sonar schools. BuPers also managed a training device laboratory in Orlando, Florida.

Chapter 4 — Full-Scale Experimentation and Development

FURTHER XN-1 AND XN-2 TESTING AND ANALYSIS

While equipment refurbishing was underway on the XN-2, sea testing continued on the XN-1.

Storage Displays

In June 1963, the testing of the first SQS-26 cathode ray tube (CRT) storage display was carried out on the XN-1. Before this, only one history of an echo return cycle could be presented at a time. With the operator unable to examine the previous history aboard ship, there was no way to compare what was detected on one sweep with what had occurred as much as a minute before on a previous one. It was hoped that the storage display now being introduced would eliminate this severe handicap in the detection process. (Computer digital storage display, of course, was not yet in existence.)

The XN-2 had initially employed a paper recorder to "store" the echo returns to the shipboard operator, but only the results from the latest ping cycle were visible. Furthermore, the paper was automatically rolled up prior to attempted reception from the next transmission. It did, however, have the advantage of creating a permanent record of echo returns for analysis purposes.

For one of the XN-2 convergence zone runs made the previous November, I had pasted together (in the laboratory) side-by-side paper traces for a sequence of 25 successive returns from a submarine crossing the convergence zone. The detection advantage of observing the past history of a target exhibiting a closing range rate became quite striking, convincing many that such a display should be developed. NUSL's John Snow made a quantitative estimate of how much the operator's threshold of detection would be improved with the simultaneous viewing of a sequence of echo returns. As a result of his studies of the XN-1 and XN-2 systems, a six-echo-return CRT storage display would be ultimately adopted in the early production systems.

Reverberation Analysis

In July 1963, NUSL's John Hanrahan completed the first quantitative analysis of XN-1 reverberation behavior. This effort would be the genesis of work leading to a thorough knowledge of the sources of reverberation, key arrival paths, and beam pattern discrimination against bottom, surface, and biological reverberation. Among other things, the investigation illustrated that a basic understanding of bottom path performance was possible, despite the reliability problems with the equipment.

Further Testing of the XN-1

In August 1963, NUSL undertook further bottom bounce echo-ranging testing with the XN-1 on *Lee*. Reporting on these tests, I noted the following "firsts":[5]

- Bottom bounce echo-ranging was achieved at 50 kiloyards on a beam aspect submarine in the baffles, 15° off dead astern, using a depression angle of only 10°. This was the first time such a shallow angle had been used, and it permitted unprecedented bottom bounce echo-ranging at 50 kiloyards, or 25 nautical miles.

- All depression angles available at that time were exercised during a beam aspect opening and closing tracking run between 45 and 10 kiloyards. Angles of 10°, 20°, 30°, and 42° were employed. This testing was the first demonstration of the possibility of holding submarine contact over the complete range of depression angles. Over the entire range interval of 35 kiloyards, there was a 75% ping return.

- Signal excesses experienced at 10° and 20° depression angles were large enough that nonbeam aspect performance should have been feasible, although time did not allow tests at other than beam aspect. Reverberation from the bottom and surface reverberation at these depression angles were found to be less of a problem than at higher angles.

Testing the Refurbished XN-2 System on Wilkinson

Sea testing on the XN-2 resumed in the October-December 1963 period in deep-water locations off the East Coast of the United States, after *Wilkinson* made a transit through the Panama Canal. Tracking performance on surface duct, bottom bounce, and convergence zone paths was examined for ranges out to 40 kiloyards. For the first time, target aspects other than beam were successfully tracked on the bottom bounce path for both low and high Dopplers.

Many problems still remained. While *tracking* performance with a narrow transmitting beam was satisfactory, the *search* performance capability over a wide sector would require further design changes. New orientations of the transmitting and receiving beam coverage were needed. Equipment reliability, while improved, was still not adequate. Despite the refurbishing effort at Long Beach, it was still necessary to make frequent adjustments to compensate for component performance drifts. Only experienced personnel carrying out continual performance checking could keep this problem under control.

OPERATIONAL EVALUATION OF THE XN-2

The decision to enter production prior to the completion of development and testing of the XN-1 and XN-2 systems redefined those systems. What were intended to be only experimental models were now *de facto* prototypes, despite their existing deficiencies. Such prototypes had to be tested by the Operational Test and Evaluation Force (OPTEVFOR), a Navy organization independent of the development community. An operational evaluation was scheduled to begin in January 1964 on the XN-2. It had been decided that the more limited XN-1 (by design) would not be evaluated by OPTEVFOR.

In retrospect, NUSL testing in the October-December 1963 period had demonstrated that the XN-2 system was not ready for an operational evaluation. Many of the identified deficiencies could not be corrected by the time the evaluation was scheduled to begin. However, after the major XN-2 refurbishment effort earlier in the year by GE, there was reluctance on the part of the SOFIX office to admit that the XN-2 was not ready.

This pressure was transmitted to NUSL's Downes, who told Washington in December that the XN-2 would be ready only if certain deficiencies were acceptable to OPTEVFOR — deficiencies that would appear to make the system incompatible with the definition of a satisfactory "prototype." Because Downes felt it was important to present the NUSL viewpoint in a positive light, he had some concern that OPTEVFOR would not understand his forewarnings. At that time,

Chapter 4 — Full-Scale Experimentation and Development

NUSL was communicating with OPTEVFOR only through the SOFIX office.

Operational Evaluation Failure and Equipment Problems Discovered

The operational evaluation began on schedule in January and continued until May, at which point it was terminated because of unsatisfactory results. In contrast to NUSL concluding that bottom bounce tracking performance was satisfactory, OPTEVFOR felt that the tracking performance was unacceptable. The consensus post-mortem was that NUSL experimental results were unable to be duplicated because the displays were seriously degraded by an electronic failure that OPTEVFOR and shipboard personnel were unable to recognize.

During the evaluation, I went aboard during an in-port visit and found the display to be performing poorly when tested with injected signals. The GE engineer who undertook the troubleshooting at my request found that a transistor had failed and needed to be replaced. How long that situation had persisted during the at-sea testing was unknown. NUSL's Harold Morrison had been aboard the ship during an earlier phase of the OPTEVFOR tests and had also reported that the display did not seem to be performing properly. In addition, he had made noise measurements that turned out to be some 7 to 8 dB higher than what was considered appropriate. The excessive noise was another problem not identified by shipboard personnel.

One important part of the evaluation was a side-by-side test against the SQS-23 on USS *Lester* (DE-1022). Alternate runs against a submarine were made with both ships using the surface duct path. It was concluded that the largely refraction-limited conditions that existed caused no significant difference for a periscope depth submarine. However, for a deep submarine, the SQS-26 showed a significant improvement in performance. Because these tests were run with the excessive noise condition, I recommended to the SOFIX office that the tests be suspended until the condition was corrected — the recommendation was accepted. Without this intervention, OPTEVFOR would have continued the testing with the system in a degraded state.

Chapter 4 — Full-Scale Experimentation and Development

Beneficial Effects of XN-2 Operational Evaluation

Even though the operational evaluation was premature, it did draw attention to the problem of turning over the XN-2 equipment in its present state to shipboard personnel who would have difficulty recognizing equipment degradation. At this point, it was clear that more attention would be required to increase performance reliability, devise better equipment-monitoring systems, and train shipboard personnel to cope with problems related to reliability and monitoring issues.

NAVY REACTION TO THE OPERATIONAL EVALUATION FAILURE

The failure of the SQS-26 (XN-2) to pass an operational evaluation received attention at the highest management levels in the Navy. The Commander-in-Chief, Atlantic (CINCLANT), wrote a letter directly to the Secretary of the Navy, Paul Nitze, with comments extremely critical of the SQS-26 program, including the statement that CINCLANT believed that the bottom bounce mode of the SQS-26 would never be an operationally viable capability. This letter again reopened the question of whether or not new construction should be equipped with the SQS-26 or with the older, but more reliable, SQS-23.

Not long before, the Navy had made a decision to go ahead with a multiyear buy of 46 ASW escorts from the *Knox* class. The initial designation of destroyer escort (DE) was changed in 1975 to frigate (FF), with the *Knox* finally designated as FF-1052. Its mission was the ASW defense of convoys, amphibious formations, and underway replenishment ships.

At the time, the *Knox* class was to be the largest class of surface combatants to be constructed in the West since World War II. It was estimated to cost some $750 million, which was considered an enormous amount of money in 1964. In year 2000 dollars, $750 million would not buy even a single *Burke* class destroyer.

SQS-26 Review Committee

The CNO response to the foregoing situation was to set up an SQS-26 review committee in May 1964 under the chairmanship of Admiral Constantine Karaberis, the head of the ASW Project Office in the Chief of Naval Material (CNM) organization. CNM had recently been organized to coordinate all Navy material bureaus. The members of the review committee included the following:

    Admiral Constantine Karaberis (CNM)
    Dr. William Carlson (TRW)
    Captain Rozier (CNO)
    Captain William Dobie (Operations, Naval, OPNAV 951)
    Commander Bradford Becken (BuShips, Code 372)
    Commander Al Glennon (OPTEVFOR)
    Dr. Dan Andrews (NEL)
    Thaddeus G. Bell (NUSL)
    J. T. Halley (BuPers)

Dr. Carlson of TRW was also a member of a special ASW committee that had been previously set up by Secretary of the Navy Nitze. Carlson's special function in the SQS-26 review, as the only member on the committee who was not a member of the Navy community, was to take an impartial view of the SQS-26. His ultimate endorsement of the soundness of the design turned out to be a major factor in keeping the SQS-26 program alive. Also present at most of the meetings was Commander John Fry of the Naval Oceanographic Office.

Commander Becken also contributed significantly to providing visibility on the advantages of the SQS-26. His recollections of the SQS-26 review committee (as included in a letter to me on 25 January 2001, in response to comments I requested on my preliminary draft of the SQS-26 history) were the following:

> The OP95 ASW Project Office had just been established under VADM Charles Martell, with its companion support organization PM4 under Connie Karaberis in NAVMAT, when COMCINCLANTFLT sent his letter of complaint on the SQS-26 not to the CNO but directly to the Secretary of the Navy, who was Paul Nitze at the time. The rumor at the time was that the SQS-26 complaint was an excuse to kill the 1052 program, a single screw ship, which the Fleet thought was a big mistake. In any case, Martell was handed the problem and tasked Karaberis to develop a response, as you well know. From my perspective at the time, the key issue raised by the Fleet was that the bottom bounce mode was not effective. Since a major ASW concern at that time was convoy escort across the Atlantic as you so well described, it occurred to me to examine

SQS-26 predicted performance along likely winter and summer convoy routes without using the bottom bounce mode. Accordingly, late one evening on my dining room table with a large sheet of paper, I laid out two such routes — a Northern great circle direct from New York to the Azores and then north to England. I used your operator's performance prediction memo to calculate detection ranges and NAVOCEANO [Naval Oceanographic Office] bathymetry and layer depth predictions. What the predictions showed was that during the winter and along the route selected typical layer depths of 1000 feet existed, providing exceptionally long direct path detection ranges. Along the summer route, good convergence zone performance could be predicted. As a result, for these important routes, whether the bottom bounce mode was effective or not was a moot point. Valid or not, the argument was seized upon by the powers that be, and I was given the opportunity to present it all the way up to Paul Nitze himself, defusing a difficult situation and providing time for you and the technical team to proceed to correct the many problems which you found in the early systems and to develop the very important rubber window.

The conclusions reached by the committee were that the SQS-26 design and performance expectations were based on sound premises, but that the project was undermanned. It recommended an expansion in the program to correct engineering deficiencies in the system, collect more oceanographic data on bottom characteristics, and improve training.

Key Decisions Following the Review

In June 1964, I attended a meeting in the Pentagon to consider further action in light of the program review. The major issue was, of course, whether or not to proceed with the SQS-26 for the *Knox* class. The Vice-Chief of Naval Operations, Admiral Claude Ricketts, presided.* I had never attended a meeting with so many flag rank officers. Vice Admiral Charles Martell, an influential supporter of the

---

*Ricketts, as a senior lieutenant on *West Virginia* in 1941, was one of the little publicized heroes of the Japanese attack on Pearl Harbor. Unlike many of his seniors, he had foreseen the possibility of an air attack and made advance plans for damage control. After six torpedoes and two bombs hit *West Virginia*, he quickly directed counter-flooding operations to prevent capsizing, saving the ship and many lives. He was promoted to full admiral in 1961. He died prematurely from heart failure at the age of 58 in July 1964, only a few weeks after our SQS-26 meeting.

SQS-26, was in attendance as OP-95, the CNO "ASW czar"; he had spent a considerable amount of time with the SQS-26 review committee. Especially remembered was the presence of Vice Admiral John S. Thach, who at the time was Vice CNO for Air.*

After presentations were given on the review committee's conclusions, Admiral Ricketts made a brief speech stating that, in the Navy's experience, pushing new technology is the way to go, even when difficulties are encountered. In his view, the SQS-26 program fell into that category and should be supported. He then asked if anyone disagreed with this perspective. The response was silence — even from the hostile CINCLANT representatives. Ricketts was such a legendary figure, especially with his four-star rank, that no one in the room was about to risk challenging his logic. Having a strong CNO supporting cast (Vice Admirals Thach and Martell, along with others of comparable stature) was also helpful.

In July, CNO informed the commanders of the Atlantic and Pacific Fleets of the various actions being taken as a result of the review. The key recommendation was that the Navy proceed with the FY64 and FY65 procurements of 27 AN/SQS-26 (CX) sonars to equip DE-1052 class ships and a shore training installation.

EXPANSION OF THE SQS-26 PROGRAM

The outcome of the review also set the course for an expansion of the SQS-26 program during the next decade. NUSL continued to perform as the Technical Development Agent (TDA), with BuShips exerting administrative control. The expanded program at NUSL was later well described by Downes:

> The work at the Laboratory in the period of expanded effort was divided into four parts. Perhaps the most intense work was in *design review*. In this effort, much overtime work was needed in order to conduct the review of each part of the whole sonar so expeditiously that

---

*An early expert and pioneer on naval air tactics, Thach became an ace at the battle of Midway, shooting down six Japanese planes. He was also former commander of Task Group Alpha, a special ASW carrier group that in the late 1950's conducted an evaluation of contemporary ASW techniques off the East Coast of the United States. I met him in 1958 during an at-sea visit, which was part of the agenda of the 1958 White Oak ASW study group of which I was a member.

the contractor would not be significantly delayed by the process. This work began midway in the design of the BX model, and continued with the changes to the XN-2 and AX models and the design of the CX model. The review was in that detail needed to be confident of the outcome. This involved matching or exceeding the competence of the contractors' engineers in vital areas. The work evolved into generating test plans for the contractor to follow in plant and barge tests. (The long-term need for barges for tests of EDO and GE systems had been presented to Washington by this time and had been met.) The success of the contractors is evident in the very notable reliability of the later SQS-26's. (Note that the Louis Allis power supply was not part of the SQS-26.*)

A second area of expanded work was the *testing* of the XN-1 and XN-2 models on the *Lee* and the *Wilkinson*. Efforts of this sort began early in 1962, and continue even now [Downes wrote these words in 1971] in the form of mutual interference investigations with the production models on other ships . . . . This [sea test] work since 1961 has successfully withstood the scrutiny of worried people in Washington. It began as the responsibility of Russell Baline and Harold Morrison, and devolved first to Frank White and then to Walter Hay. Testing aboard the *Wilkinson* occupied 7 of the last 8 years of her commissioned life. Tests were planned in great detail, and executed on schedule, despite many problems that could have caused delays.

The third area of work was in *checking out the production SQS-26's after they were installed*. This important but unglamorous work was begun under Walter Wainwright. With the passage of time, the scope of the work changed to include participation in various tests involved in the acceptance of new ships, and, thereafter, in assistance to ships and Fleet commands. This has required a very considerable amount of travel to meet various needs, sometimes with very short notice indeed.

The fourth area has been that work aimed *at equipping the Fleet to use the SQS-26's effectively*. This has been a task that has expanded the role of the Laboratory. Beginning as furnishing guidance in how to

---

*It was unfortunate that the responsibility for the Louis Allis power supply (LAPS) ended up outside the sonar code at BuShips because it was a troublesome item that seriously affected the reliability of the SQS-26 system. I recall being at sea on a 1052 class ship in the 1970's witnessing an ASW exercise in the Mediterranean. When the LAPS stopped working, the ship's ASW operations shut down completely, with the crew unable to determine the cause. By the time that the expert on LAPS arrived in the Mediterranean from NAVSEA (formerly BuShips), the sonar had been inoperative for some 10 days. It took less than an hour to repair the system by replacement of a defective transistor.

set various system controls to accommodate a particular ocean environment encountered, it has evolved into guiding oceanographic data gathering, suggesting tactics, participating in sea tests and exercises, and instructing Fleet personnel. This work became the responsibility of Thad Bell.[6]

Conduct of the Expanded SQS-26 Development Program

I was to be heavily involved in the systems engineering for two of the four areas of program expansion described above by Downes: (1) testing the XN-1 and XN-2 models on *Lee* and *Wilkinson* and (2) equipping the Fleet to use the SQS-26's effectively. Testing of the XN-1 and XN-2 at sea was, of course, essential for providing input to all other parts of the expanded program.

As noted earlier, the decision had been made to concentrate on developing the XN-2 into a prototype model. As a result, the XN-1 mission was relegated to one that involved only selected testing.

Expanded XN-2 Development and Testing

In June 1964, a new development assist project was set up (D/S 331) to cover another year of equipment modification and sea test work on the XN-2. This project, which would involve the nine sea tests that were to be conducted between July 1964 and May 1965, had two broad objectives:

- Providing the information necessary to redesign the *Wilkinson* system into a prototype that would serve as a model for modifying production system specifications and hardware.

- Developing guidelines that could be used to train shipboard operators in optimizing equipment settings and in predicting performance once those settings were made.

When NUSL first went to sea for the XN-2 testing, the priority was such that it was reasonable to assume the ship and submarine would be employed around the clock. However, it was soon discovered that this approach was not a productive use of time at sea. There were only a limited number of knowledgeable personnel, none of whom could be expected to work without sleep. Everyone was finally convinced that a 12-hour day was all that most participants could efficiently handle.

When testing was not being conducted, the time was spent checking the equipment, carrying out repairs, digesting and analyzing the data from the previous period, and planning tests for the next period. The actual workday was about 16 hours.

NUSL's Walter Hay, the technical test director (TTD), performed the invaluable function of handling the details of running the tests and ensuring that the equipment was properly operating, while I concentrated on how well the test results were conforming to expectations and what new tests should be run. Keeping the operation running smoothly and spending the at-sea time wisely seemed to require both our efforts.

The time spent at sea was optimized in the following manner. After it was determined that about 12 days should be the maximum continuous period at sea, the group would leave port on a Monday, begin 10 days of testing with the submarine at the location of interest on Tuesday, and then head back a week from the following Thursday. (This schedule allowed the ship to have a day in port (Friday) before the weekend.) When the ship returned to port after the 10 days of testing, the data would be carefully analyzed, the results and future plans would be discussed with other knowledgeable personnel, the proper experts would be lined up to participate in the next set of tests, and the equipment repairs, modifications, or special dockside tests would be made. About 3 weeks in port were required to complete these activities before the next sea trip would begin. This was the typical 5-week cycle during sea tests on the two SQS-26 experimental systems.

Developing a Prototype Design

The systems that would require retrofitting the improvements made in the evolving "prototype" system included the AX systems already installed; the BX systems, all in production; and the CX systems on the verge of going into production. While 1 year was available to further refine prototype specifications based on XN-2 testing, it was clear that this time must be carefully spent.

Identifying a problem and its likely fix resulted in the following actions: (1) temporary modifications to the hardware, (2) testing of those modifications, (3) incorporation of any further needed modifications, (3) conduct and analysis of new tests on the second set of modifications, and, finally, (4) preparation of specifications for the permanent change.

This process meant that the XN-2 modification and testing programs could not be too ambitious if all was to be completed within the year. On the other hand, the end result of these efforts required solid specifications for use as the basis of changes to the characteristics of the AX, BX, and CX systems.

Work continued on improving equipment reliability, especially on the serious problem of drifts in the component characteristics. It was still the era of analog circuits, which were very sensitive to temperature, vibration, and aging, so that there was a continuing struggle to maintain satisfactory performance in this area. To solve the frequency drift problem meant converting some of the reference frequency circuits to a digital design. The component performance drifts causing the most trouble were found in the new storage displays and in the matched-filter receivers. The testing at sea was backed up by recording signals prior to operation of the matched-filter receivers and then by playing these signals back for examination at shore-based activities, such as NUSL, Tracor, and GE. The capability for comparing both existing and alternative processing and display techniques using the recorded at-sea data in a laboratory setting was invaluable. In addition, key component voltages and frequencies could be checked to determine whether or not the shipboard system had been performing as designed. It was not unusual that sources of equipment problems were first revealed during the analysis of these recordings by Tracor.

Providing Guidance in Equipment Operation

To provide guidance for shipboard operators in the use of SQS-26 equipment meant developing models of equipment performance in the sea environment. This approach required an understanding of both the environmental effects and the manner in which the equipment would react to those effects. Thus, at any given time and location, it was necessary to be able to estimate the propagation loss to a potential target, background levels, and the effect of the medium on the distortion of transmitted signals in their travel to and from the target.

The performance modeling development was difficult not only because of the infinite combinations of environmental characteristics in worldwide naval operations, but also because all the environmental characteristics of the sea that had a significant effect on sonar performance had not yet been identified. While it was known that bottom loss was important, it was not understood why sea water attenuation seemed to vary with location. It was not even completely certain that attenuation really did vary with location or whether observational errors resulted in misleading conclusions.

Conveying what was being learned about optimizing equipment settings and predicting performance to the XN-2 operators initially required the preparation of lectures for *Wilkinson's* crew (lecture notes were also distributed to other ships receiving SQS-26 installations). While this material presented the fundamentals of how the system was expected to work in a sea environment, it was inadequate as an instruction for two reasons. First of all, calculation methods required to make decisions on how the equipment should be operated were too complex to be used by typical shipboard personnel. Secondly, even if the calculations could be simplified, the knowledge of key environmental input characteristics was lacking. For example, not enough was known about the medium in any particular location to make an accurate and timely estimate of *in situ* propagation loss. This situation was not only true for the bottom reflection path, but for the convergence zone path as well. More about providing guidance for the operator will be discussed in chapter 8.

General supporting investigations initiated under the expanded D/S 331 development program, but not closely coordinated with equipment development and testing on *Wilkinson*, will be covered in chapter 6, Supporting Research and Development.

Fault Recognition

In January 1965, a fault recognition system that would allow frequent checks on the following key indicators was set up to address the problem of unrecognized malfunctions:

- Transmitter frequency,
- Minimum detectable level of injected test echo,
- Beam direction (receiving and transmitting),
- Background level,
- Transmitter voltage and phasing, and
- Driver pulse shape.

An estimate was made that this approach would improve the percentage of time that nontechnical personnel were able to recognize subtle faults from 50% to 90%. For example, the beam direction indicator often revealed that the real beam was going to a bearing that was 5° different from the ordered bearing.

Gulf of Mexico Tests

Previous testing in the North Atlantic had been confined to locations off the East Coast of the United States with water depths from 2700 to 3000 fathoms. In January 1965, it was decided to try bottom bounce experimentation in the Gulf of Mexico. Here, the water depth would be only 1900 fathoms with a sound speed profile not permitting convergence zone formation. As shown in figure 8, in a location such as this without a convergence zone, a similar high-intensity convergence of sound can occur within the bottom reflection field, given a negative thermal gradient at the sonar. In this environment, a minimum occurs in the bottom reflection propagation loss between 34 and 38 kiloyards, depending on depth.[7]

Another expected advantage of the Gulf of Mexico environment was the lower bottom path reverberation from sea surface backscattering occurring at angles in the vicinity of 5°. This effect was observed only for those downward refraction conditions that did not allow a surface duct path to the target. However, with the deep isothermal layers commonly found in that location, the reverberation from the strong ducting interfered with the bottom bounce reception and also weakened the bottom bounce sonar field. Although, in actual operations, the surface duct path from the deep duct would be useful for detection, this was not what was being tested. In addition, the absence of bottom loss measurements in the area created uncertainties in performance expectations. *In situ* measurements resulted in differences in bottom loss from one part of the area to another that amounted to as much as 7 dB one way and 14 dB to the target and back.

Finally, a biological reverberation problem was encountered that showed a time-of-day dependence, as indicated in figure 9. It was surprising to find the essentially identical dependence on time of day that Dr. Robert P. Chapman had observed north of Bermuda with explosive

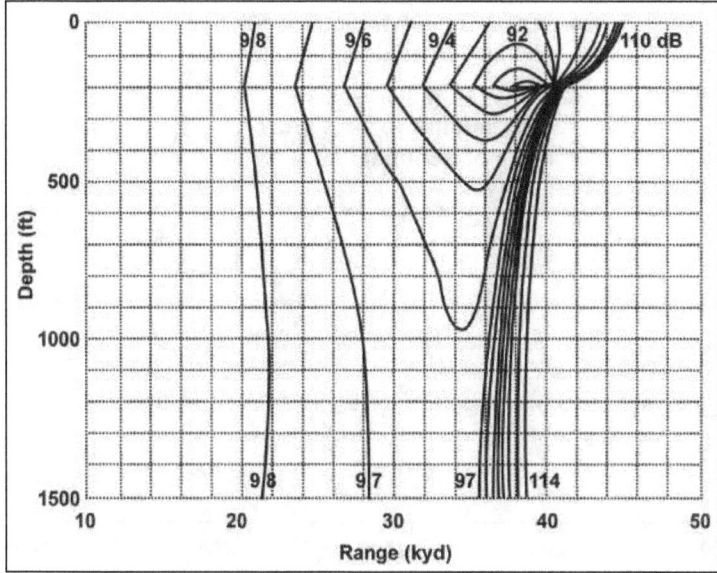

Note: Maximum intensity of bottom-reflected sound falls between ranges of 34 and 38 kiloyards, depending on depth.

Figure 8. High-Intensity Zone Occurring in Bottom-Reflected Sound When Bottom Depth Is Not Sufficient for Convergence Zone Formation

measurements. Later, the same time-of-day dependence for biological reverberation was found in the Mediterranean Sea.

Before it was finally recognized that reverberation was increasing systematically each day, this effect had caused difficulty in the interpretation of performance versus submarine aspect. Because no time-of-day effect was expected, a routine had been established for carefully determining the accuracy of equipment performance by beginning the day with beam aspect echo-ranging, where there would be no doubt regarding the target location or strength of the echo. Careful measurements of propagation loss, minimum detectable signal, and echo level were taken to determine system capability. With this information and an estimate of target strength decline with aspect away from beam, a prediction was made of how echo-ranging performance would change with target aspect. The prediction would indicate that quite adequate performance was to be expected at the finer aspects. As the day went on,

Chapter 4 — Full-Scale Experimentation and Development

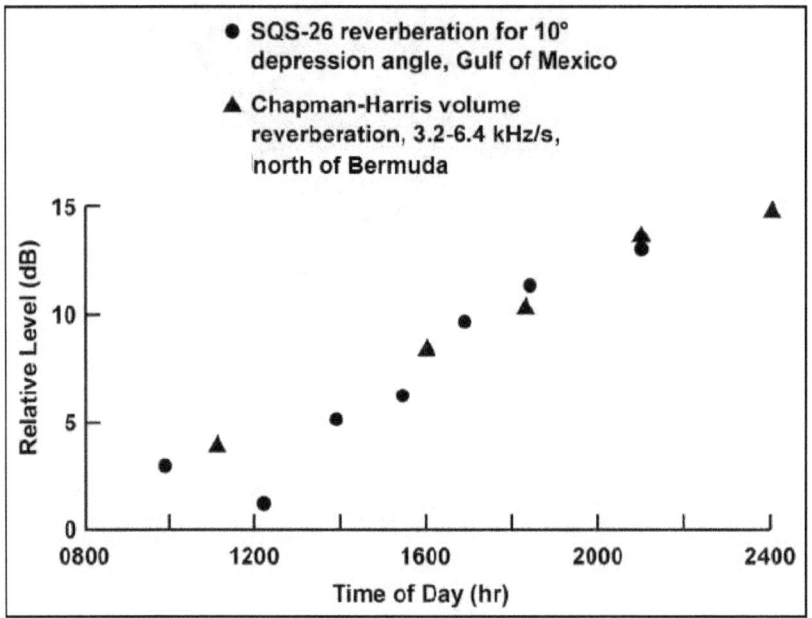

Note: Also observed by Robert Chapman north of Bermuda in biological backscattering measurements with explosive sources.

Figure 9. Systematic Dependence on Time of Day in the Gulf of Mexico Due To Biological Reverberation

successively finer aspects would be tried. By the late afternoon when the most difficult worst-case geometry bow aspect was finally attempted, the performance was consistently disappointing in comparison with predictions made earlier in the day.

Initial investigations were made of the variables that could be involved in performance changes, such as propagation loss, wind speed as it might affect surface backscattering, own ship background noise, display/processing degradations, and thermal conditions. Only belatedly was it discovered that the reverberation was systematically increasing with time of day. What seemed to be an abnormal decay in performance at the finer aspects was only a matter of the reverberation increasing as the aspect gradually departed during the day from the beam-on geometry. Although it may seem surprising that this behavior was not discovered much sooner, it must be remembered that all those involved were fully occupied during the 12-hour testing day in ensuring that all tasks were

Chapter 4 — Full-Scale Experimentation and Development

carried out as planned and that the equipment was free of subtle malfunctions. Such unforeseen correlations can create confusion in the interpretation of experimental results.

Wide-Sector Transmission

During the January testing, a 120° search sector was employed for the first time in bottom bounce echo-ranging operations. Previously, the search sector had been limited to the original design of 30° per step. No serious problems were noted with the wide coverage from display crowding, clutter, or excessive reverberation. The advantage of wide-angle coverage was evident when the operator was able to achieve detections without previous knowledge of the target bearing.

Signal Processing

By February 1965, a number of alternative signal processing schemes had been tested. Linear processing in place of clipped processing seemed to offer a slight improvement, about 1 to 2 dB, depending on the signal-to-background ratio. Analysis by GE in the laboratory indicated that for nonbeam aspect targets the longer echo return benefited from the use of a longer time constant at the output of the correlator. However, this result did not include the effect of the scan-converter storage display, which in the scan converter process already provided a long averaging time to the correlator output. Comparisons of display receptions with the examination of the signals at the display input revealed no obvious degradation in display capability from the loss of resolution in the scan-converter storage process. Although there was confidence that no large degradations existed in the pulse-compression processing and display, in some ways the results were below expectations. For this reason, it was always hoped that some major source of performance loss (which could be readily fixed to provide stronger performance) would be found.

In October 1965, Tracor published a classic summary of their signal processing and display studies that had been directed by Augustus (Gus) F. Wittenborn under BuShips contract Nobsr-93140. Perhaps the most important conclusion was that "... for bandwidths of up to 400 [Hz] ... for the energy received within a single resolution interval, no correlation loss exists."[8] The loss, often thought to be the result of the "correlation"

Chapter 4 — Full-Scale Experimentation and Development

process, was now attributed to the spreading of the echo energy from some combination of Doppler and time spreading outside the ideal resolution interval of the waveform. The principles relating to this loss had been discussed in a landmark paper published earlier in the year by Weston.[9] Stewart later introduced an apt term when he called it "energy splitting loss."[10]

Performance Prediction Under Test Conditions

The inability to find any major problems in signal processing implementation was reflected in performance prediction success during testing. Basic measurements of propagation loss, source level, and minimum detectable level against an injected signal were good indicators of what could be expected in echo-ranging performance, despite early concerns that pulse-compression correlation processing would not hold up against real-world, bottom-reflected echoes. This observation, of course, does not imply that the performance prediction problem would be solved under operational conditions, where there would be no opportunity to measure propagation loss to the target, no knowledge of target aspect, and no experts available for calculating the predictions.

Displays

The scan-converter-type storage display, although valuable for presenting the past history of as many as 12 pings, was acquiring a reputation for perhaps the most troublesome hardware component in the system. Instability, nonuniformity over the area presented, and adjustment complexity were continuing headaches. Unfortunately, the display problems persisted well after the prototype had been developed. As late as November 1969, NUSL's Downes wrote a special memo on display problems, urging more effort on a corrective program. The importance attached to this problem was indicated by his statement that ". . . the need to do these tasks seems to be paramount among all other SQS-26 needs at this time."[11]

Shallow-Water Performance

Echo-ranging performance in shallow water (commonly defined as water with a depth of less than 100 fathoms) with predecessor sonar systems had been a problem area for years, with poor propagation and high reverberation produced by interactions with the bottom. Although conventional wisdom assumed that the SQS-26 would have the same difficulty, NEL had already experimented successfully with echo-ranging using the low-frequency LORAD system in a number of shallow-water locations. I had discussed the LORAD results with NEL's Mackenzie, who was quite optimistic about the prospects for long-range, low-frequency echo-ranging in shallow water. Bottom backscattering, while expected to produce the dominant reverberation background, would be minimized by low grazing angles at the long ranges expected for low frequencies. With the results of the Colossus propagation measurement program conducted by NUSL in the late 1950's, it was already known that propagation loss would be better than commonly expected.[12]

In May 1965, an echo-ranging experiment was conducted in a typical shallow-water area (30 fathoms) south of Long Island, New York — a location that had been measured earlier during the Colossus program. Although a severe negative thermal gradient with downward refraction was encountered, the bottom loss for the sand bottom typical of shallow water was only a little more than 1 dB per bounce, as expected. This condition permitted good propagation with multiple bounces out to the target and back. At the low angles of incidence involved, the reverberation was also reasonably low. In the late 1940's, I had conducted experiments with the QHB in the same location under downward refraction conditions. For those experiments, the detection ranges were limited to less than 1 mile so that multiple bounce ranging was not possible.

With the completion of the May testing in shallow water, project D/S 331 ended, and the specifications for the prototype were "frozen." The extent of the required hardware changes was such that it would take nearly 2 years, despite the priorities on the program, to incorporate them before taking the prototype to sea on *Wilkinson*. The new model would be dubbed the SQS-26 (XN-2 MRF) for "major retrofit." In the meantime, the production systems being manufactured for the DE-1052 class were to be built to the new specifications.

Performance Improvement with Time During Development Testing

Improvements in the equipment made during the XN-2 testing showed up in the test results. In the initial technical evaluation from November 1962 through March 1963 under T/S 25, it was possible to demonstrate bottom bounce performance only for a beam aspect submarine. However, under the follow-on D/S 281 project from October 1963 through December 1963, bottom bounce echoes were obtained on a nonbeam aspect submarine — but only for wind speeds below 10 knots. During project D/S 331 in1964 and 1965, the performance improvement trend continued.

A statistical compilation was made of results from the first seven of nine D/S 331 trips from July 1964 through March 1965. Out of 563 pings in wind speeds ranging from below 10 knots through more than 20 knots, 450 *nonbeam aspect* echoes were received via the bottom bounce path for an overall 80% echo-to-ping ratio.

CHAPTER 5
PROTOTYPE TESTING

TESTING THE SQS-26 (BX) PRODUCTION SYSTEM

The BX, considered the first fully capable SQS-26 production model to reach the Fleet, was tested on USS *Wainwright* (DLG-28) in deep waters of the North Atlantic off the coast of Florida. While the BX system did not completely benefit from the D/S 331 tests completed in May 1965 and the major retrofit (MRF) specifications to be used for all GE production systems, many of the improvements that resulted from that earlier testing were incorporated. Performance comparable to that demonstrated on USS *Wilkinson* (DL-5) was demonstrated for shallow-water, surface duct, and convergence zone sound paths. However, performance was less satisfactory for bottom bounce echo-ranging because of instability in the signal processing.

During the BX tests on *Wainwright* in June 1966, Captain William Peale, the SQS-26 program manager in Washington, DC, was on board to observe the testing. Expressing concern about the operator's obvious lack of knowledge concerning the fundamentals of setting up the equipment in any given environment, he inquired about the willingness of NUSL personnel to visit the Fleet ASW School in San Diego, where they would (1) provide guidance (via lectures) in those areas of knowledge in which operators were weak and (2) attempt to define other areas where a problem might arise.

In response to Captain Peale's request, Richard Chapman of NUSL arranged for a 1-week series of lectures and discussions at the Fleet ASW School, beginning on 22 August 1966. As previously noted, lectures, while always well received, are no substitute for a formal course of study. The training problem will be discussed further in chapter 8.

TESTING THE XN-2 MAJOR RETROFIT PROTOTYPE

By November 1966, the MRF version of the SQS-26 (XN-2) had been installed on *Wilkinson*. Between then and August 1967 when the system was finally ready for sea testing, GE and NUSL were busy conducting dockside checkouts, identifying problem areas, and making the appropriate fixes. As with any new design, many changes were made

Chapter 5 — Prototype Testing

to the system during the debugging process. A lengthy test plan had been prepared and was carefully followed to ensure that the prototype system would attain the specified dockside performance characteristics before it was taken to sea.

The first shallow-water sea test in August 1967 under T/S 51 revealed that there were still some problems. Despite this, NUSL's test director Walter Hay noted that the new MRF version showed significant improvements in stability and reliability over the original (unretrofitted) one. In successive months of testing, performance substantially improved as problems were discovered and then corrected. With over 160 system deficiencies recorded, a considerable portion of the total sea test time was still devoted to the correction of hardware problems. Hay noted that such efforts were not unusual for new equipment that was as complex as the SQS-26 MRF hardware.

One example of the type of situation encountered with the "improved" system involved the performance of the new digital FM sweep generator. The digital version had been provided to overcome the very troublesome FM waveform frequency instability problem caused by the previous analog design. However, the initial generator introduced harmonics outside the design frequency band, causing reverberation interference with the CW waveform in an adjacent band. The issue was resolved only after considerable time had been diverted to carefully examining what was happening, devising a fix, and then installing and carefully testing the modification.

By the end of August 1968, seven sea tests (totaling 54 days) had been conducted against a submarine target by *Wilkinson* with the MRF version of the SQS-26 (XN-2). Such intensive testing was reasonable for a system that would serve as the prototype model for the new CX production system and for the retrofitting of changes in the AX system (eventually known as the AXR).

In October 1968, presentations of the test results on the MRF were made to representatives of the Naval Ship Systems Command, the ASW Systems Project Office, CNO, the Secretary of the Navy, and the Secretary of Defense for Research and Engineering. In December, NUSL (Hay and I) presented the same results to the Commander, Operational Test and Evaluation Force (COMOPTEVFOR) Headquarters, in Norfolk, Virginia. The major objective of the presentation was to summarize key results in anticipation of COMOPTEVOR conducting an operational appraisal of the *Wilkinson* system. As it turned out, the appraisal was conducted instead for an early SQS-26 (CX) production model on USS *W. S. Sims* (DE-1059).

During the presentation, system reliability was shown for sea tests three through seven (after problems from the new MRF installation had largely been brought under control). The mean time between failure (MTBF), omitting the transmitter system, was 145 hours, as compared with a goal of 100 hours. The transmitter system was excluded because the system still in use aboard *Wilkinson* represented the original design that consisted of two transmitter drivers feeding the beamforming elements in each of the eight layers of the transducer array. This design was replaced in production by a solid-state module drive for each of the 576 elements of the array, which resulted in a far more satisfactory performance. For example, with the new modular transmitter design (along with other changes), the MTBF for the entire CX system, after the first 3 years of operation, was 500 hours. In contrast, the transmitter design used on *Wilkinson* during testing constituted a major reliability problem.

The mean time to repair for the *Wilkinson* system was 1.7 hours, compared to a goal of 1.5 hours (again excluding the transmitting system). With the CX, the goal of 1.5 hours would be attained for the whole system.

Wilkinson's MRF availability (excluding the transmitter) was 97.4%, compared to a goal of 98.5%. The availability for the entire CX would be 99.7%.

A second area of concern for *Wilkinson* involved the self-noise that resulted from the current procedure of not painting the steel dome surface because of problems with paint deterioration. A by-product of this approach, unfortunately, was either surface corrosion or biological fouling, both of which required periodic grooming by divers.

Another subtler self-noise problem was caused by line components of the transmitter B+ power supply that could be coupled through the transmitter output transformers to the receiver circuitry. When one of the lines was picked up by a CW comb filter, the high-Doppler CW receiver would be jammed. This situation was temporarily solved during the technical evaluation by changing the frequency of the 400-Hertz generator to 390 Hertz, thereby moving the bothersome harmonic out of the CW band. It should be noted that this problem was the result of *Wilkinson's* unique transmitting system.

Finally, a third noise source was generated when the ship's engineering personnel were shifting the operating load on various pieces of machinery. For example, operations such as crossconnecting the main boilers increased the noise level for substantial periods of time before steady state operating conditions were again reached.

The presentation to OPTEVFOR describing the tests results compared them to the Specific Operational Requirements that had been established prior to the testing. The test results were consistent with those requirements. Where problems were encountered, fixes were provided.

Thus ended SQS-26 (XN-2) testing and development aboard *Wilkinson* for the 6 years from mid-1962 through mid-1968. Although it was never envisioned that this amount of time would be required for the SQS-26 development and prototyping stages, it did not, in retrospect, appear to be of unreasonable duration for a system that had introduced so many innovative techniques.

UNRECOGNIZED FAILURES IN THE PRODUCTION EQUIPMENT: THE SEA TEAMS

The foregoing reliability statistics on equipment largely addressed failures that shipboard personnel readily *recognized*, thus permitting the effect of these failures on system availability to be quantified.

More difficult to address are *unrecognized* failures, which might not be identified for months, thus degrading equipment performance without anyone aboard ship realizing that such a situation exists. It has already been mentioned that such failures caused problems in the first operational evaluation, which led to Fleet personnel attempting to maintain equipment in peak condition without any training or the benefit of experienced NUSL engineers. At that time, only the NUSL engineers were able to recognize subtle equipment degradations.

Chapter 5 — Prototype Testing

An initial attempt to address the production system problems resulted in the establishment of sonar evaluation and assistance (SEA) teams.[1] These teams, composed of experienced engineers and technicians, were able to go aboard ship, determine where the unrecognized degradations existed, and then feed this information to the design community for the development of corrective measures (where feasible). In the near term, the SEA team would attempt to fix the problem, as well as explain to shipboard operators how it could be recognized.

Under the SEA team program, two three-man teams — one on the East Coast and the other on the West Coast — visited SQS-26 ships on a schedule established by NUSC and the type commanders. These 4-day dockside visits were carried out on 93 ships between July 1970 and September 1973, until such time as British Rear Admiral Hill put it so aptly "when active sonar research and development tended to take a back seat [to passive sonar]."[2]

Figure 10 shows the number and type of problems encountered by the SEA teams, along with the percentage of ships encountering each type. It can be seen that the SEA team encountered display deficiencies producing serious degradation on 40% of the ships visited. These display problems were especially troublesome because they were difficult for the ship's force to recognize.

Not included in the SEA team visits (confined to equipment) was self-noise (reported on separately). Noise was a chronic problem for the

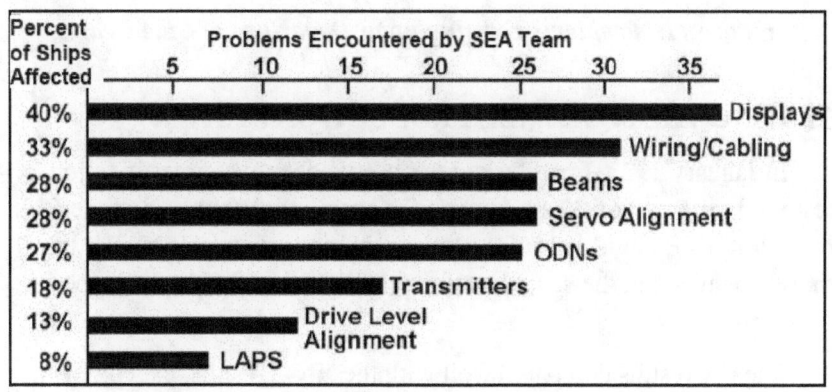

Figure 10. Problems Encountered by SEA Teams

Chapter 5 — Prototype Testing

predominant steel dome of that era, and it accounted for a source of sonar system degradation even more severe than that from the displays. Furthermore, ship's force was unable to readily recognize noise deficiencies.

Although the SEA teams fixed what problems they could, some required either more time or more specialized attention than the teams could provide. Figure 11 shows the number of problems that still existed after the SEA teams departed — again, the displays were the leading offenders.

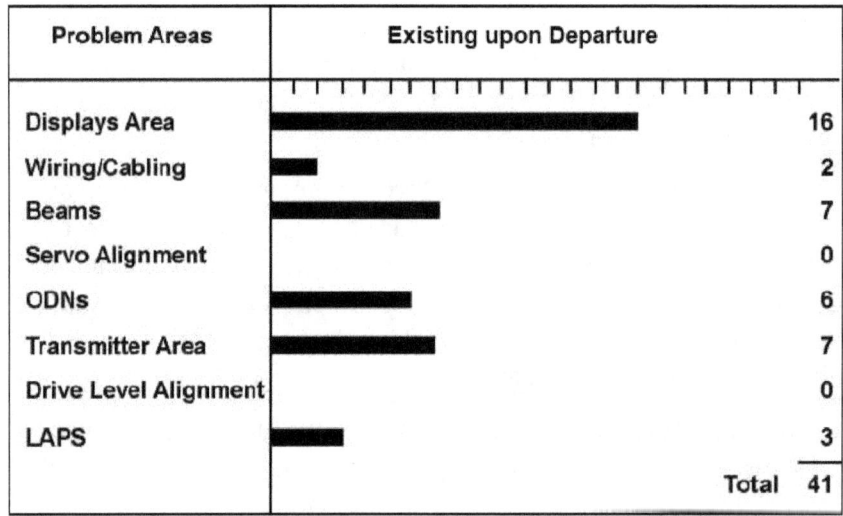

Figure 11. Problems Existing upon Departure of SEA Teams

TESTING THE SQS-26 (CX) PRODUCTION SYSTEM

In January 1971, I was tasked to observe deep-water certification testing during part of the first cruise for operational appraisal of the CX by COMOPTEVFOR on *Sims*. Runs were made that were similar to those conducted in the same location with the XN-2 MRF prototype on *Wilkinson*.

The CX results compared well with the XN-2 results, except for unsatisfactory CW Doppler processing. This situation was unfortunate because the operational appraisal results, presented by OPTEVFOR as characteristic of a CX model, showed that the Doppler processing was still ineffective. Yet, later convergence zone tests on the *Connole* SQS-26 (CX) — observed by this author — showed the type of favorable CW Doppler detection performance that had been seen on *Wilkinson*.*

It should be noted that a problem with the CW receiver had shown up in underway signal differential tests with an injected signal before echo-ranging testing began;† this deficiency should have been repaired prior to the *Sims* operational appraisal. Moreover, the CW performance on *Sims* actually had been normal during the dockside testing with injected signals. It is also possible that the CW processing results were the result of a pit-log input problem at sea that was unrecognized by shipboard personnel.

Figure 12 shows the seven major subsystems of the SQS-26 (CX) components as they existed in the early 1970's: (1) transmitter, (2) power distribution, (3) array, (4) test and monitoring, (5) receiver, (6) program and control, and (7) display.

---

*NUSC personnel, who were isolated from the *Sims* operational appraisal beyond the first cruise, presumably so as not to exert any influence on the testing or the results, did not see the final report until a year after the tests had been completed. At that time, they wondered why OPTEVFOR did not question the unsatisfactory CW results since NUSC had previously provided OPTEVFOR with satisfactory *Wilkinson* results under similar conditions.

†These tests should be performed by ship's force at least once a quarter.

Chapter 5 — Prototype Testing

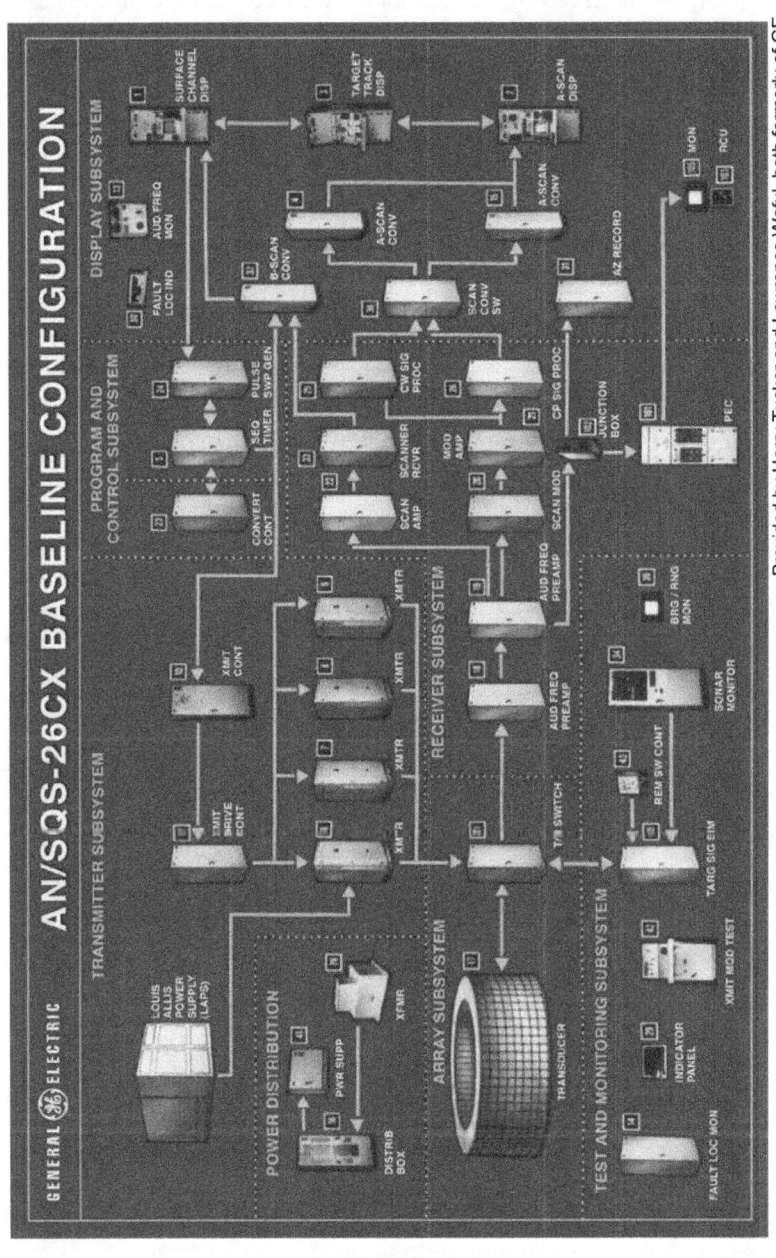

Figure 12. Components of the SQS-26 (CX) in the Early 1970's

CHAPTER 6
SUPPORTING RESEARCH AND DEVELOPMENT

Supporting research and development work was performed from May 1964 through May 1965 under the expanded D/S 331 charter, with a portion of this effort actually initiated prior to the start of D/S 331. A number of these special projects, pursued apart from SQS-26 equipment hardware development and testing, are described below.

TRACOR CONTRACT

To increase analytical manpower for the conduct of studies in support of the SQS-26 sonar, the SOFIX office in Washington, DC, contracted with Tracor Inc. in Austin, Texas. In spite of the coordination problems that resulted from the Tracor contract being managed directly out of the SOFIX office and from the physical distance existing between New London and Austin, NUSL welcomed the additional research talent brought to bear on SQS-26 problems.

For several years, Tracor, under the direction of Dr. Augustus (Gus) F. Wittenborn, performed much valuable research for the SQS-26 program, including those efforts regarding signal processing that were already mentioned in chapter 4. Other projects managed by Tracor under the SOFIX contract are described below.

Dome Water Effects on Bearing Accuracy

In late 1963, Tracor's G. T. Kemp conducted an early study on bearing accuracy as influenced by the phase distortion from sound speed changes attributed to temperature fluctuations in the array dome and dome water. Although the effect on sound speed due to the heating of the dome water during the sound transmission process had to be considered, the bearing errors that it produced did not appear to be serious.

Effect of Nonworking Elements in the Array

On 21 December 1964, Kemp completed a study of the impact on beamforming from nonworking elements in the array. This information was used to determine when inoperative elements should be replaced.

Chapter 6 — Supporting Research and Development

Display Studies at Tracor

In May 1966, James Young of Tracor published the first of what was to be a series of valuable experimental studies on the effect of display marking level, signal-to-noise ratio, false alarm rate, and the number of signal levels presented.[1] In a September report, Young presented his conclusions:

- Intensity-only modulation was nearly as effective as intensity-plus-deflection modulation. Both types were initially provided with the CRT displays in the SQS-26 systems, but the required number of side-by-side traces on the operational displays made deflection modulation impractical. With marking from multiple targets, it became especially confusing (in viewing a history) to know what trace to associate with a given mark. A target from one trace could obliterate a target on an earlier trace. It was important to learn from the Tracor investigation that little was sacrificed in providing intensity-only modulation, even in the idealized situation of a single target with nonoverlapping traces.

- A minimum marking density of about 0.2 is required for up to three quantized signal marking levels. An even higher marking density may be beneficial when there are more marking levels.[2] (A human can distinguish among approximately seven intensity marking levels.)

- Approximately 50 hours of detection training is required to bring an observer with no experience up to the capability of an experienced observer.

The Tracor experiments assumed perfect normalization (i.e., sufficient control of signal and background levels to maintain them within the dynamic range of the display). In practice, the continuous variation in the reverberation background, the nonuniform display characteristics over the face of the CRT, and the drifts in adjustments with time made the ideal test conditions of the Tracor simulation difficult to achieve aboard ship. In fact, NUSL's primary objective aboard ship was to maintain dynamic range in the display to the extent that a target signal would appear different in level from background interference. This characteristic had to be obtained over all parts of the display, over the passage of time, and for both reverberation and noise.

Dome Water Problems

Because of the apparent errors (as much as several degrees) previously observed in XN-1 beam depression angles, studies were made by Tracor regarding the refraction effects caused by fresh water in the sonar dome, with the results of this research reported in June 1966 by Kemp.[3]

At that time, fresh water was used in the dome to avoid the contamination potentially present in sea water, especially when the ship was close to port. Tracor investigators concluded that while some errors in beam depression angle could be caused by refraction due to fresh water, these errors would not amount to more than 1°. Later, however, the concern about bearing errors of as much as 1° resulted in replacement of the fresh water in the dome with sea water from deep-water locations.

BIOLOGICAL REVERBERATION

In May 1964, the Canadian Dr. Robert P. Chapman (affiliated with the Naval Research Establishment (NRE) in Dartmouth, Nova Scotia) presented a paper on volume reverberation in the North Atlantic at the *22nd U.S. Navy Symposium on Underwater Acoustics* at NUSL.[4] With Harris, Chapman had already written a classic paper on sea surface backscattering[5] that contained a highly useful empirical formula for its prediction, given the angle of incidence, frequency, and wind speed. It was later found that the formula contained in that paper would permit NUSL to predict SQS-26 surface reverberation with good accuracy.

From discussions with Chapman after the 1964 symposium, NUSL learned that volume reverberation caused by backscattering from the gas bladders of fish would depend on the time of day and the operating frequency. In the SQS-26 frequency region, backscattering levels were comparable to those expected from a 20-knot wind speed at a grazing angle with a surface of 15°.[6] Chapman eventually continued his experimental work in biological backscattering, covering most of the world's strategic oceans and their adjacent seas.[7]

In March 1966, Backus and Hersey of Woods Hole Oceanographic Institution (WHOI) completed a study on the expected geographic dependence of the volume reverberation experienced by the SQS-26. They postulated that the intensity of biological reverberation should be related to the density of fish. In an earlier seminal paper, they had presented evidence that the swim bladders of fish were responsible for the deep-ocean resonant backscattering effect evident in acoustic observations on the continental rise south of New England.[8]

Backus and Hersey further postulated that the fish density would be proportional to the density of available plankton, the food upon which the fish depended for subsistence. A plot of plankton density would thus provide a rough prediction of what to expect in backscattering strength. A few "spot" observations of biological backscattering in various locations suggested that their correlation was a reasonable one. This research was not only immediately useful in planning further experimentation, but it was also ultimately expected to be another valuable input into decisions regarding the choice of shipping routes during ASW threats.[9]

ARRAY RECEIVING PHASES

In October 1964, Richard F. Sweetman of GE made receiving phase measurements on the SQS-26 barge array at New York's Lake Cayuga testing facility, with a source positioned 200 feet away. He found that the measured phases were consistent with theoretical expectations, indicating that — in the receiving situation at least — *there were no significant interelement coupling effects.* Thus, receiving beamforming networks could be designed on the basis of the acoustic dimensions of the array being equal to the physical dimensions.

DISPLAY RESOLUTION

On 9 December 1964, Boivin and Thorp reported measurements on the range resolution of the new display, which used a scan converter tube for storage. The resolution of the display was about 65 milliseconds, as compared with the 10-millisecond, pulse-compressed output of the FM processor.[10] Although there was an initial concern — due to a 10-millisecond, pulse-compressed echo being presented against a reverberation background increased by a ratio of about 65 to 10 — the scan converter resolution, in retrospect, appeared to provide a net benefit in the presence of the typical echo elongation caused by both the medium and the target reflection process.

MARINE GEOPHYSICAL SURVEY PROGRAM

To permit accurate estimates of propagation loss via the bottom bounce path, the largest single effort resulting from the 1964 program review was initiated — the Marine Geophysical Survey (MGS). The MGS program measured bottom characteristics, mainly bottom loss, along the critical sea routes to Europe; it would later address routes to Hawaii in the Pacific Ocean and routes within the Mediterranean Sea.

Under the auspices of the Naval Oceanographic Office (Wilburt Geddes would become manager), the MGS program was in effect from 1965 through 1970.

Managing the MGS Program

The contract to conduct the surveys and analyze the data so that useful information was obtained for operation of the SQS-26 system was assigned in May 1965 (as the result of a competitive process) to two contractors: Alpine Geophysical Associates and Texas Instruments.

There was a considerable amount of criticism from the scientific community regarding the MGS approach to acquiring information on the ocean bottom. It was generally felt that the Naval Oceanographic Office did not have the depth of scientific expertise to oversee such a program, while the nonprofit oceanographic research organizations possessed the expertise required to gather the new data. However, because it was indeed more of a *survey* effort than a *research* program, the Navy selected the Naval Oceanographic Office as the management activity, which, as it turned out, was a good decision.

The MGS effort was so unprecedented, however, that this author also had early concerns. Studying the acoustic properties of the deep-ocean bottom on a mass scale, where the basic information was 1 to 3 miles below the ocean surface, was a formidable effort. While acoustics had been used in deep water since the 1930's to study the nature of the earth's crust, there had been little interest in the acoustic properties of the ocean bottom as they might influence submarine detection. As a matter of fact, through the late 1940's, "the deep ocean" had been defined in the sonar school manuals as "water so deep that the bottom played no part in acoustic propagation." The bottom was considered to be important to sonar only in water depths less than about 100 fathoms. It was not until 1947 that Robert Young of NEL presented the first experimental evidence in the open literature that bottom reflection might indeed be an important contributor to sound propagation between near-surface points in deep water.[11]

The contractors participating in the MGS program were well-known and respected scientists with many years of experience. Dr. Charles Officer, the author of a book on underwater acoustics, headed the Alpine effort and the reputable Dr. Claude Horton of the University of Texas served as the Chief Consultant for Texas Instruments.

Although the Naval Oceanographic Office did not provide recognized experts in acoustics, this agency was thoroughly experienced in ocean surveys and operated under a naval staff that was extremely responsive to Navy needs. Management of the program was excellent, especially the sensitivity to NUSL requirements.

Role Played by NUSL

NUSL's primary objective was to develop techniques from the MGS results that the Fleet could use to estimate bottom bounce performance. Acting in this capacity, NUSL was also responsible for two other functions: (1) informing the Naval Oceanographic Office regarding the types of information that should be gathered and (2) providing guidance to that Office on measurement techniques and associated analysis methods.

This author was part of a Naval Oceanographic Office steering committee — along with Robert Urick and Robert P. Chapman (among others) — that oversaw the MGS program as it progressed. With program reviews held at frequent intervals to discuss results and progress, a member of my staff, John Hanrahan, was enlisted to provide assistance in reviewing the details of the ongoing effort.

Processing the Explosive Data

In January 1965, Joseph Collins at Tracor completed a study on how to process the MGS bottom loss data. Because pulse-compression processing was used with the FM waveform for bottom bounce echo-ranging in the sonar system, a tie-in between that processing and the survey measurements made with explosive sources was required. Collins determined that processing the explosive data with a 10-millisecond averaging time would provide the information on what bottom loss the FM pulse-compressed waveform would experience.

It turned out that in later years NUSC had second thoughts about this approach, although at the time it seemed completely reasonable. After further studies were made in the 1980's, it was decided that a total energy measurement would have made more sense than the measurement of returns with a 10-millisecond resolution.[12]

Concerns About Measurement Accuracy

Early in the survey program, NUSL reviewed the measurement techniques used by both Alpine and Texas Instruments. Before the survey program was underway, a "spot" check in April 1965 was completed on the Alpine measurement methods, but there had been no opportunity to do the same for Texas Instruments. There was some concern that the Texas Instruments calibration methods might be less accurate than the methods employed by NUSL and Alpine.

The best way to determine whether or not the Alpine and Texas Instruments calibration methods were consistent would be to have the two contractors make measurements in the same locations. While this approach would involve expense beyond that planned, it would ensure that all the measurements were accurate. The result was the occupation by the two contractors of a sizeable number of stations in the Norwegian Sea during the summer of 1966. The exercise, informally dubbed the "Turkey Shoot" (or, alternatively, the "Summer Festival"), showed that the results from both contractors compared well, despite the differences in their measurement methods. Thereafter, throughout the program, arrangements were made whereby one contractor would revisit stations that had already been occupied by the other in an effort to maintain quality control. With this approach, it was found that the accuracy of the measurements was excellent.

Some suggested that NUSL could have avoided this situation by equipping each contractor with identical measurement and analysis setups. Although this may seem to be a good idea, "standard" setups often have hidden problems that are never discovered, and differences in the interpretation of measurement results can affect the standard way of performing what are complex processes. Moreover, there was much more accomplished than a simple comparison of the measurements of one contractor with the other. These measurements were also weighed against those of other activities, either by having an activity visit one of the measurement stations or by locating some of the stations in areas where other activities had earlier made bottom loss measurements.

An Unfortunate Diversion

One lesson — learned the hard way — was to be skeptical of including nonessential measurement items in the plans. For example, a thermistor chain was suspended from a buoy so that the temperature versus depth profile could be measured down to a few hundred meters at a fixed location as a function of time. While this was not part of the survey objectives, it was hard to argue against a presumably negligible cost addition that would provide information about temperature changes with time in the upper layer of the ocean. The buoy would be picked up within a few days of its planting.

As it turned out, use of the buoy had a significant impact on the cost of the ships. First, the radar mast had to be built high enough to detect the buoy. Next, a large structure had to be provided on the fantail of each ship to launch and retrieve the buoy, which was much larger than originally envisioned. Finally, there was a considerable amount of valuable ship time expended to find the buoy, particularly in adverse seas. Needless to say, the use of the buoy was discontinued after these problems were recognized.

MGS Survey: Cost and Grouping

On a budget of some $25 million, the survey was to cover 20 task areas in the key shipping lanes of the North Atlantic, North Pacific, and Mediterranean Sea. The size of the average task area was 350,000 square miles, with the calculated cost of the bottom loss survey at approximately $3.50 per square mile — a surprisingly small cost for the amount of information to be acquired. It was fortunate that NUSL's Eugene Podeszwa had found that the bottom loss measurements could be grouped into rather large homogeneous domains, which could be ranked in accordance with their loss — nine significantly different domains were finally identified.

Comparison of MGS and SQS-26 Measurements

In January 1968, a comparison was made between the MGS bottom loss chart for areas off the East Coast of the United States and the propagation loss measurements made during the course of SQS-26 testing for that month. The charts showed the mean bottom loss in an MGS "province" as a function of grazing angle.

During the XN-2 (MRF) testing, two provinces were sampled over a ship's track of about 400 miles, with average losses in each province not significantly different from those predicted in the corresponding MGS charts. This experience further confirmed the logic of using propagation loss measurements taken with explosives for performance predictions of SQS-26 waveforms. Secondly, it validated the reasonableness of constructing bottom loss domains by grouping measurements from a large number of discrete locations.

ATTENUATION COEFFICIENT

A March 1965 analysis of previous convergence zone propagation measurements for SQS-26 operations in both the Pacific and Atlantic suggested an increase in attenuation from the Pacific to the Atlantic by as much as a factor of two. A few years later, attenuation was found to be still greater in the Mediterranean Sea. This was a complete mystery at the time because it was assumed that the attenuation at any given frequency would be identical in all sea water, regardless of location. It would take another 10 years before Mellen and Browning of NUSC explained that this condition was due to a location-dependent pH and a relaxation effect from boric acid.[13]

SHIPBOARD PREDICTION METHODOLOGY

On 7 December 1965, I presented the first formal paper on converting oceanographic and sonar information into SQS-26 performance predictions at the *Navy-NSIA Oceanography Workshop* held at NRL.[14] For the bottom bounce mode, NUSL would generate tables of detection probability in the coverage annulus, given inputs of (1) performance figure (source level minus noise level), (2) depression angle, (3) water depth, and (4) location. All values would be computed in the laboratory because digital computers suitable for performing the task aboard ship were not yet available. Even if a computer had been accessible, the basic environmental information on deep sound speed, bottom loss, and biological backscattering was not. It would be another 10 years before NUSL personnel would assemble the data and develop the techniques and computer hardware to provide shipboard computations of predicted ranges and optimum equipment settings for the major strategic oceans and seas of interest. As part of the process of gathering basic input information, Podeszwa generated, in meticulous fashion, the necessary atlases of sound speed versus depth and location that became standard references for computations of propagation loss in the deep ocean.[15]

In August 1969, NUSL's Richard Chapman made observations of SQS-26 performance during the first CX ASW training exercise. Although the detection performance was generally satisfactory, Chapman found higher than average reverberation levels, which resulted in performance that was poorer than what was predicted in the latest available NUSL sonar performance prediction manual. After returning to New London, Chapman proposed that a shipboard computer design be initiated to measure the reverberation levels *in situ* so that this information could be used to calculate a performance prediction rather than having shipboard personnel use nominal reverberation levels and precomputed performance tables, as was the current practice. The first minicomputer, introduced by DEC in 1965, had not yet been used to solve the shipboard performance prediction problem. NUSL's George Brown was assigned to investigate the development of such a shipboard computer device and, as soon as funding could be found for the latest DEC computer (the PDP-11), work would begin.

In November 1972, Brown issued a summary report on an experimental shipboard performance prediction computer that became the prototype for the first model of the sonar *in-situ* mode assessment system (SIMAS), which was eventually added to all SQS-26 sonars. The early models were primitive by today's standards, using tape for stored data input instead of a disk drive and paper printouts instead of the CRT display that has become the standard today, even in home computers.

Other organizations proposed the use of larger computers at central locations, with the results of each prediction communicated to the SQS-26 ships. A number of such systems were developed, produced, and put into practice. While such computers could indeed make use of more elaborate computation programs, the accuracy of their predictions was seriously limited by the lack of up-to-date input information on the shipboard environment (e.g., thermal conditions, background noise, and strength of reverberation). These central computations had to depend upon estimates of *average* conditions at the ship locations. The significant difference between actual conditions and statistical averages was the very problem that led to the NUSL design of a *shipboard* reverberation monitor and computer system in the first place.

SQS-26 DISPLAY TESTING AT NUSL

In November 1967, Herbert Fridge and Peter Cable performed an experimental study on an actual SQS-26 CRT display set up in the laboratory at NUSL to determine the relative advantage of presenting from one to six echo histories on the display. They performed a total of 1,788 measurements with six Navy sonar technicians. It was concluded that for the same false alarm rate the detection threshold was lowered by 3 log of the number of histories presented. This result was different from one previously obtained by Tracor, which showed a 7-log improvement with the number of echo histories presented. Although a complete rationale for the difference was not developed, it appeared to be related to the greater uncertainty presented to the operator for the NUSL tests with respect to target range, bearing, and range rate.

SCATTERING STRENGTHS IN SQS-26 TEST AREA "B"

In January 1969, an analysis of bottom-scattering strength measurements from four sources was completed in SQS-26 test area "B," which is a 1° square that is 700 miles off the east coast of Florida. The measurements were obtained from three versions of the SQS-26 (XN-1, BX, and AXR) and an Admiralty Underwater Weapons Establishment (AUWE) system. In April, an analysis was also made of XN-2 bottom-scattering measurements in the same location. It was found that all five sets of measurements were consistent with the "Mackenzie model," which was based on bottom-scattering measurements made off the West Coast of the United States with the LORAD system.

Chapter 6 — Supporting Research and Development

JOINT OCEANOGRAPHIC ACOUSTIC AND SYSTEM TESTS

The Joint Oceanographic Acoustic and System Test (JOAST) Program, conducted in the Mediterranean Sea in the late summer of 1970, was an unprecedented applied research effort for developing the modeling information necessary to evaluate the expected performance of an operational system over a wide variety of environmental conditions in an area of vital importance to the Navy. NUSC's Bernard Cole organized this program based on insights he acquired during his early participation in the SQS-26 exploratory tests in the Mediterranean environment.

The effort — involving a research ship (*Sands*) and an operational ship (USS *Glover* (AGDE-1)) — permitted experimentally based inferences of what to expect from SQS-26 performance on long-range paths in the Mediterranean given observations of the environment. The following relationships were established:

- Biological scattering strength *and* reverberation in the convergence zone;

- Convergence zone propagation loss *and* the total sound speed profile, depth of receiver, and operating frequency;

- Bottom characteristics *and* bottom path echo-ranging performance against a target submarine over the whole deep-water Mediterranean basin.

CHAPTER 7
THE RUBBER DOME WINDOW

STEEL DOME PROBLEMS

Maintaining the paint on the steel domes of both USS *Willis A. Lee* (DL-4) and USS *Wilkinson* (DL-5) in good condition was a major concern. As the paint deteriorated, rough surfaces were produced in the near term and corrosion and marine fouling in the long term, all of which caused serious noise problems at operational speeds.

In July 1962, NUSL's William Downes made an analysis of paint erosion on the steel bow domes housing the SQS-26 (XN-1) on *Lee* and the SQS-23 on USS *Randolph* (CVS-15).[1] He noted that the loss of paint was related to the inner framework provided for structural reinforcement of the dome. One mechanism that especially seemed to cause problems involved the flexural motion of the dome face that occurred between welds on the framework.

The paint problems were serious enough for the Navy to establish an interim policy of going to sea with unpainted dome windows. Although this approach minimized the near-term noise problems, the dome surfaces became corroded and fouled with marine growth if they were not scrubbed every few months.

RUBBER DOME WINDOW PROPOSAL

Goodrich Contract

In February 1963, the Navy issued contract 89483 for the development of a pressurized rubber dome* that had been proposed by the B. F. Goodrich Company as an alternative to the existing SQS-26 steel dome. This design evolved into a rubber window that was inset into an otherwise steel structure. The bottom part of this dome was steel, and a flexible steel cord was embedded in the rubber to provide sufficient

---

*The rubber dome window (RDW) was adopted as the standard descriptive term for the Goodrich design. Unfortunately, the term can be confusing at first encounter because the adjective "rubber" applies to the window, not to the dome. "Rubber window dome" would have been a more appropriate description.

window strength. Internal water pressure in excess of that outside the window was used to maintain the dome's window shape. No longer required was the rigid steel rib structure that had been used to reinforce the steel dome window. An antifouling chemical that slowly leeched out of the rubber window kept the surface clear of marine organisms.

Preliminary Testing of the Rubber Dome Window

The SQS-26 RDW was first used on *Lee* during the period from July 1965 to April 1966. Measurements made in August 1966 indicated an improvement in noise levels of 6 dB below 12 knots and 3 dB above 20 knots. Between 12 and 20 knots, the gain slowly decreased as speed increased.

In September 1969, Julius Natwick of NUSL reported on an experiment to determine transmission loss, indicating that little or no loss occurred on a 1.25-inch wire-reinforced rubber panel with material typical of that used in the dome window.[2]

Bradley (DE-1041) Rubber Dome Window Installation

Between 17 November 1971 and 20 March 1972, a second RDW was installed on USS *Bradley* (DE-1041). On 31 July 1972, NUSC's Savas Anthopolos, Jr., conducted a series of self-noise measurements as a function of both ship speed and Prairie Masker condition (Off or On). The Prairie Masker was an air bubble screen that "masked" the noise coming from the ship's propeller and machinery spaces. Over a speed range from 12 to 28 knots, the RDW with the Prairie Masker in the "On" condition showed an incredible improvement, averaging about 15 dB over the standard that was set for the steel dome window.[3] The greater advantage seen on *Bradley*, as compared with *Lee*, was attributed to *Bradley* being a quieter ship due to the Prairie Masker system.

Rubber Dome Window Noise Measurements Compared with Sea State Ambient Noise Calculations

A perplexing effect observed in the *Bradley* measurements was also seen in later measurements on subsequent RDW installations. In the previous installations of the steel dome window, it was customary to see the computed level for expected sea state ambient noise on the same graph where the measured noise versus speed values were shown. This convenient reference would indicate how far the measured levels were above the ambient sea noise levels in cases where the measured levels were the result of a combination of propellers, machinery, hull vibration, and flow noise.

What appeared unusual to those viewing the RDW measurement curves was that at the low speeds the measured noise was often as much as 6 dB lower than the calculated ambient noise. The first reaction was that there must be some error in the measurements because the measured combination of ambient and ship noise sources should clearly not be less in magnitude than the ambient noise alone.

The problem was that the ambient calculations, based on measurements made with *omnidirectional* hydrophones, were corrected for the directivity index of the sonar. The directivity index was applied to the omnidirectional measurements on the assumption that the noise was approximately isotropic, that is, nondirectional. However, the actual sea noise originated from the sea surface, and it was uncertain how directional this noise appeared to the sonar.

The solution to this puzzle in which the RDW measurements were lower than sea state noise was not provided until a decade later in an example worked out by Burdic in his widely used sonar analysis book.[4] Using a vertical array with nearly the same wavelength dimensions as the SQS-26, Burdic calculated the sea surface ambient noise received by the array for a sine-squared dependence of the radiated noise on vertical angle. For the vertical beam of the array steered to the horizontal direction, the calculated sea surface noise received was 6 dB lower than that computed with the directivity index assumption for array gain.

The 6-dB difference between the actual array gain and the directivity index in ambient noise calculations was just the error that had been observed when SQS-26 ambient noise measurements were compared to those calculated with the directivity index. After this discovery, 6 dB were added to the directivity index in the computation of sea state ambient levels. This capability for providing accurate estimates of the effect of sea state noise on sonar noise was especially important for the RDW installations, where low ship noise often meant that ambient noise would become a significant contributor to overall noise levels.

Chapter 7 — The Rubber Dome Window

ECHO-RANGING PERFORMANCE WITH RUBBER AND STEEL DOME WINDOWS

In late 1972, a consensus of those involved was reached on the need to conduct an experimental demonstration regarding the operational value of the SQS-26 RDW before an expensive program was initiated for installing these rubber windows on all SQS-26 ships. In January 1973, CNO's Rear Admiral Jeffrey Metzel (OP-981) directed COMOPTEVFOR and the Naval Ship Systems Command (NAVSHIPS) to proceed with plans for testing the effect of the RDW on echo-ranging performance. The responsibility for the technical planning and conduct of the tests would later be assigned to NUSC.*

As a result of the planning, side-by-side tests were scheduled between USS *Knox* (DE-1052) with a rubber dome window and USS *Kirk* (DE-1087) with a standard steel dome window. The target submarine was USS *Guitarro* (SSN 665). The location of the testing was north of the Hawaiian Islands, along the 157th west meridian between latitudes 27° and 33° North.

In accordance with CNO direction, the main objective of the tests was to compare echo-ranging performance in the convergence zone. The convergence zone ranges for the described location and time of year were about 30 miles. To ensure that biologic reverberation would not adversely affect the testing, arrangements were made for Norbert Fisch from the NUSC research department to conduct volume-scattering measurements a month before the testing by dropping explosive sound sources and sonobuoys from a P3B.

Knox and *Kirk* were carefully groomed so that both sonars would be in excellent operating condition. However, shortly before the tests were to begin, a 7-inch cut in the rubber window on *Knox* had been discovered. Underwater repair was successfully accomplished with a special bonding material (Concresive) supplied by B. F. Goodrich. Further noise measurements showed that the dome had been restored to

---

*At the working level, Walter Hay and I worked out the details, with outstanding cooperation from all the NUSC staff (including Technical Director Harold Nash and Commanding Officer Captain Milton McFarland), as well as from CNO OP-981 (Rear Admiral Jeffrey Metzel and Captain Thomas Glancey) and NAVSEA.

its original condition. No other dome problems occurred during the remainder of the test period.

For the first 7 days of testing, the speeds were generally restricted to 15 knots because the ship had limited fuel. I had explained to the OPTEVFOR representative that the noise versus speed curve of *Knox* was such that the noise at 22 knots was the same as it was at 15 knots, which, in theory, meant that the convergence zone echo-ranging performance at 22 knots would be the same as it was at 15 knots. The concept was difficult to accept because steel dome noise behavior had shown 22-knot noise levels that were 20 dB higher than those at 15 knots, making convergence zone echo-ranging unthinkable.

The OPTEVFOR representative reacted as follows to the possibility of a high-speed convergence zone search with a rubber window 1052 class: After refueling, he asked, why not use speeds of 22 knots instead of 15 knots if the *Knox* performance would be just as good? Although I agreed (somewhat hastily) to use the 22-knot speed for *Knox*, it was decided to run *Kirk* at speeds of 12 to 15 knots to maximize data acquisition. This approach would still result in an approximate 11-dB improvement in noise level. The *Knox* decision was actually not quite so daring as it appears because one convergence zone run at 22 knots had already been made with no evident degradation in performance. This earlier result, however, was based on only one sample.

Comparing the performance of *Knox* at speeds of 22 knots with *Kirk* at speeds of 12 to 15 knots showed the following improvements in convergence zone performance: (1) the *Knox* zone width was increased by a factor of four and (2) the echo-to-ping ratio over the *Knox* zone was 80% as compared with 50% for the smaller *Kirk* zones.[*] These results, of course, were obtained with *Knox* at a 7- to 10-knot higher speed than *Kirk*.

---

[*]NUSL's Frank White, who had spent many years attempting to improve SQS-26 performance by devising modifications to the sonar transmitters and receiver processing, reacted by stating that this solid piece of mechanical engineering — the rubber dome window — was far more effective in increasing sonar performance than all the other electronic improvements taken together.

Chapter 7 — The Rubber Dome Window

EXPECTED OPERATIONAL IMPACT OF THE RUBBER DOME WINDOW

In late 1973 through early 1974, NUSC undertook a study of alternative surface ship sonar concepts — one was based on the current system with its rubber dome window and improved signal processing techniques that were about to be introduced.[5]

It was decided to model system performance in a task force protection scenario involving a transit from Norfolk, Virginia, to Gibraltar. As described in chapter 4, Commander Bradford Becken had suggested something similar in the 1965 CNO committee review of the SQS-26 program for the steel dome system. However, nearly a decade later, the advantage of the rubber dome window, along with much more quantitative data about system performance and bottom loss characteristics, would be available.

Figure 13 shows the selected track across the Atlantic. November was chosen as a representative month of the year in terms of wind speed and thermal layer depth. Next, 30 stations were selected along the track, with 120-mile spacing between stations, so that calculations could be made for representative variations along the route. These stations, each with known statistical distributions in wind speed and layer depth for November, permitted the selection of random samples that simulated the variability of the environment.

Figure 14 illustrates a notional escort formation, with the spacing based on lessons learned from earlier exercises with SQS-26 ships. Although the spacing assumed the availability of 35-mile convergence zones, it would be adjusted as required in the parts of the route where no convergence zones existed because of insufficient water depth.

Table 1 shows (1) the number of stations on which the search path gave the best performance, (2) the equipment mode, and (3) the mean detection range. The deep-water operating guidelines for the SQS-26 systems directed a search in the convergence zone mode if the water depth and thermal conditions allowed. Otherwise, the search was conducted in the bottom bounce mode (if bottom loss and wind speed permitted). If neither bottom bounce nor convergence zone operations were possible, the surface duct would be searched with the processed directional transmission (PDT) mode. In shallow water, the PDT mode

Chapter 7 — The Rubber Dome Window

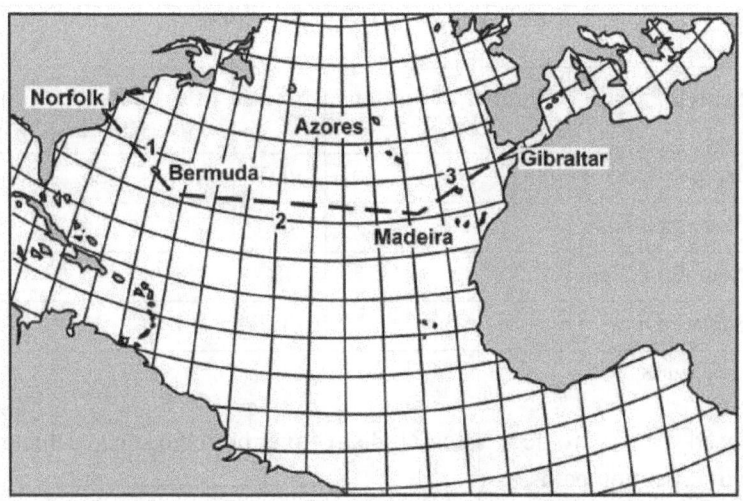

Figure 13. Three-Segment Norfolk to Gibraltar Route Selected for Computations of SQS-26 Detection Performance

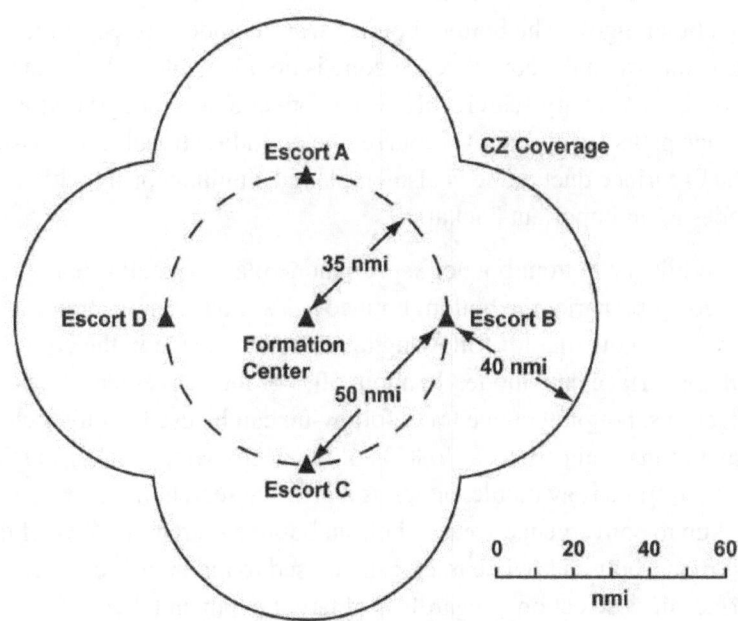

Note: The formation would be adjusted in conditions unsuitable for convergence zone coverage.

Figure 14. Notional Escort Spacing for Four SQS-26 Ships

Table 1. Detection Range Statistics Computed for Environment on Norfolk to Gibraltar Route

Search Path	Number of Stations	Mode	Mean Detection Range (Kiloyards)
Shallow Water	2	PDT	25
Convergence Zone	21	CZ	86
Bottom Reflection	4	BB*	25
Surface Duct	3	PDT	26

*Bottom bounce

was used with a zero depression angle to cover both the surface duct and bottom reflection paths.

The reason for emphasizing the convergence zone mode in deep water in the training programs is fairly obvious from the above results. In many locations, the convergence zone mode is not only the most reliable long-range path in deep water, but it also allows the largest detection ranges. The bottom bounce search mode is relegated to those occasions when the convergence zone is not available. The surface duct is used as a primary search only when convergence zone or bottom bounce paths are absent. Of course, the omnidirectional transmission (ODT) surface duct mode is also employed simultaneously with the other modes as an important backup.

While the bottom bounce *search* mode plays a small role in the foregoing scenario, the bottom bounce *track* mode demonstrates a significant potential for following up detections made in the convergence zone or surface duct mode. In about 50% of the convergence zone detections, bottom bounce track follow-up can be used to either close the target or maintain contact. In 82% of the deep-water stations, *surface duct* coverage is available, either as a primary search mode or as a backup to convergence zone or bottom bounce search. In 60% of those situations, bottom bounce track can be used to maintain a contact on surface duct detections, regardless of target depth and thermal conditions.

The original role of the bottom bounce mode was envisioned as providing a search capability that would be independent of thermal gradients and target depth. It now appears that the most important role for the bottom bounce mode in many locations is to provide a track follow-up option to convergence zone and surface duct detections

There are important caveats regarding the above conclusions. Coverage in the convergence zone requires a combination of surface temperature and water depth that is favorable to its path formation. In some deep-water locations, the surface temperature may be too high to permit the satisfactory formation of a convergence zone along an entire route. In this event, the role of bottom bounce search will tend to become more important than it was in the above scenario. The same conclusion would apply in intermediate water depths, say 100 to 1000 fathoms, where convergence zone coverage is often not found, regardless of surface temperatures.

An important advantage of the convergence zone path is that the determination of whether or not this propagation path is available can be obtained from a simple slide rule designed by NUSC's Eugene Podeszwa, given ocean basin, surface temperature, and water depth inputs.[6] The slide rule also provides an accurate estimate of the range to the leading edge of the zone.

Although a depression angle of 5° will normally offer good coverage of the convergence zone, beyond this, the ship must be concerned about the reverberation produced by (1) biologics in some locations and seasons during certain hours of the day and (2) surface backscattering in high winds.

The availability of the bottom bounce path, on the other hand, requires information on ocean bottom reflectivity, which is available only in selected locations. The geometrical coverage of the bottom path is dependent on water depth and depression angle, which must be obtained from charts or computations. The usability of the path for a given range and depression angle is also quite sensitive to wind-speed-dependent reverberation. Furthermore, the bottom bounce path shares with the convergence zone and surface duct paths the same sensitivity to biological reverberation.

CHAPTER 8
EQUIPMENT OPERATION AND TACTICAL EMPLOYMENT

EARLY CONCERN ABOUT OPERATOR TRAINING

The 14E12

Shipboard training problems had been highlighted as a significant concern during the May 1964 SQS-26 review. One response by the Navy was to initiate the procurement of a shipboard recorder and playback system, which would be known as the 14E12. Such a device was appealing because it could feed real echoes into the sonar that could then be displayed to show the operator how a submarine echo would actually look. However, there was a failure to recognize that playing back echoes under some specific condition offered no help for the two primary training needs: (1) setting up the system in any of the many various environments and (2) predicting what performance could be expected in that environment.

NUSL Requirements Study

In February 1966, after completing a study of shipboard training requirements for SQS-26 sonar systems, NUSL's Fridge defined the four areas in which training would be necessary:

1. Basic operator training in the use of sonar controls;

2. Operator training in recognizing echoes and in distinguishing submarine from nonsubmarine echoes;

3. Team training for the entire ship watch section in the detection, tracking, and attack of a submarine target;

4. Training of sonar supervisors, ASW officers, and the ship's commanding officer in methods for employing the ship and its sonar in an ASW operation, in various environments, and in tactical situations.

The conduct of (1) and (3) aboard ship, in the absence of a real submarine target, necessitated shipboard hardware capable of injecting target echoes with levels dependent on a particular environment, as well as injecting submarine range, course, speed, and depth. The conduct of (2) required the high fidelity that was present only in recorded information and which, in practice, was incompatible with the flexibility required for (1) and (3). Accomplishment of (4) was best achieved with suitable shore-based classroom courses of study. Attempting to satisfy the four requirements resulted in a number of fundamental problems.

First of all, providing a full shipboard target simulation capability for both operator and team training in a system as complex as the SQS-26 required more than a supplementary black box. The degree of integration needed with the system operation in reality necessitated undertaking this task at the same time that the system was being designed, rather than as a modification effort. If such an approach were not taken, training expenses would become prohibitive.

Secondly, training in the recognition of echoes required equipping at least some ships with an elaborate recording capability and then exposing these ships to both real and false targets in a number of operating environments. The recording had to be conducted by a team of experts able to properly operate the recording equipment, use the appropriate sonar modes, and identify the nonsubmarine targets. For playback, all ships had to be equipped with a suitable playback capability and instruction material so that an operator could understand what was being seen. If this could not be done, the playback had to be carried out in a shore-based training facility.

Thirdly, providing guidelines for the optimum employment of the ship in any operating environment presupposed that such knowledge existed somewhere and that this knowledge could be made available for classroom training. However, fulfilling this requirement would be difficult in the early stages of system employment before the operating techniques for the various environments were fully developed.

Finally, BuPers (Navy) and the Fleet were responsible for training material and the conduct of training programs. Coordination between these activities and the personnel conducting the SQS-26 procurement program would be especially important.

The only specific near-term actions that resulted from the NUSL training study are described next.

Chapter 8 — Equipment Operation and Tactical Employment

- NUSL prepared and distributed documents that described how to use the built-in test set on each system for training an operator on equipment settings with a variable-intensity moving target. However, there was never any indication that shipboard personnel used these training guides. With all the demands on the time of personnel at sea for watch standing and readiness drills, shipboard sonar training did not receive high priority, at least during peacetime. Furthermore, the operators were not graded on their capability to operate the sonar, which lessened their incentive to spend time on such skills.

- NUSL generated technical documents on employing the equipment properly at both watch supervisor and command levels. As previously noted, in February 1964, this author presented a series of lectures, along with lecture notes, to the crew of USS *Wilkinson* (DL-5), with the hope of providing a capability for use of the SQS-26 during the operational evaluation when engineers were not available to offer personal guidance.

- In April 1966, a final version of the above notes — in the format of a shipboard manual to guide equipment operators and command levels on the operation and employment of the SQS-26 for submarine detection — was published for general distribution. As it turned out, the publication was too complex for use by shipboard personnel and lacked much of the basic input information on the environment that was still in the process of being acquired (bottom reflectivity information, for one example). It did, however, present a core of information on fundamentals that was essential to the development of future methodology for SQS-26 operating manuals.

- By early 1969, four NUSL reports were issued, each containing information that was tailored to the AX, AXR, BX, or CX production system. These documents were easier to use and contained more environmental information, although they still were not fully adequate for the typical shipboard operator.

NUSL gave many lectures at Fleet commands, to sonar squadrons, and to the sonar schools. Although these efforts were well received, they fell considerably short of providing shipboard personnel with the expertise required to properly employ the sonar. What happened was that NUSL personnel just did not have sufficient time to furnish the necessary instruction. During World War II, design engineers were often commandeered, once the shooting started, to spend as much of their time as required on training. In peacetime, however, the priority of design engineers, particularly those working for an equipment acquisition command, was to *develop* equipment rather than to spend time in training personnel on its use.

With the formal closing of development and testing for the *Wilkinson* prototype at the end of 1968, it was becoming more and more difficult for NUSL to expend the effort that would ensure shipboard operators were properly trained in equipment use. Although the overall SQS-26 budget at NUSL reached a peak of $3,756,000 in FY67 (July 1967 through June 1968), supporting 85.6 man-years of effort, it thereafter declined. In FY68 (July 1968 through June 1969), planned funding from NAVSEA (formerly BuShips) was especially short in the training area.

EXPANDING THE SQS-26 OPERATING DOCTRINE PROGRAM AT NUSL

Those responsible for the operation of the new ships were still very much concerned about the training problem. Early in 1969, delivery of the large procurement of 48 SQS-26 (CX) equipments to the Fleet began. Training in the use of those systems now had to become a priority matter, with the funding problem properly addressed. Accordingly, in March 1969, Rear Admiral Leslie J. O'Brien, now Director of the ASW & Ocean Surveillance Division at CNO, called a training conference.

CNO Training Conference and Its Impact

Rear Admiral O'Brien was the same Commander O'Brien introduced in chapter 3 who had played a key role 14 years earlier in promoting the SQS-26 concept. The outcome of the CNO conference was a rejuvenated operating doctrine program at NUSL, with NAVSHIPS directed to provide increased funding specifically earmarked for operating doctrine development. Richard Chapman, who by then was working on my staff, would manage this effort, with the assistance of Juergen Keil. Special attention would be given to the following areas:

- Maintaining close contact with the operating Fleet so that lessons learned were incorporated into the new SQS-26 publications;

- Taking on the responsibility of ensuring that all official Fleet publications would be kept current with regard to SQS-26 doctrine;

- Incorporating the latest scientific information (such as bottom loss, biological reverberation, bottom scattering, and convergence zone phenomena) into the new publications;

- Providing assistance to the various Navy training programs; and

- Separating the operating doctrine efforts into operator-level and command-level guidance.

Setting up Training Areas at Sea

An important early effort (1970) in the expanded doctrine and training program directed by Chapman and Keil was the assistance provided to the Fleet in setting up at-sea training areas in the Caribbean near Guantanamo Bay and in the Pacific off San Diego and Hawaii. Not only were locations selected where the sonar conditions would be suitable for the convergence zone and bottom bounce modes of the SQS-26, but procedures for detection and tracking runs against submarines (or surface ships simulating submarines) were worked out.

To prepare ships for ASW exercises, tactics were prescribed for multiple ship sweep and contact holding operations. After Fleet approval, these tactics were published in the Naval Warfare Publications series that formed the basis of Fleet operating doctrine. Manuals were generated by NUSC for each version of the SQS-26 equipment at both the equipment operator and command level.

Fleet Feedback Program

In an especially valuable effort, Chapman and Keil devised a system for obtaining written feedback from each SQS-26 ship participating in an ASW exercise. At the request of the Fleet, the results of this feedback were carefully analyzed in the NUSC laboratory.

Feedback was also obtained from NUSC personnel participating as observers and advisors on Fleet ASW exercises in all the world's oceans.

Chapter 8 — Equipment Operation and Tactical Employment

More about the results from the Fleet feedback program will be provided in chapter 9.

Navy Shift in Emphasis to Passive Sonar

The operating doctrine program for the SQS-26 continued as described above from 1970 through 1974. After 1974, the combination of towed array techniques and a large population of noisy Soviet nuclear submarines meant that passive sonar techniques would become predominant in the U.S. Navy. As a consequence, training in active sonar methods was not stressed.

The new emphasis on passive sonar did not occur overnight, but gradually increased during the early 1970's, beginning with an experimental towed array being installed on USS *Patterson* (DE-1061). In 1973, passive sonar employed in an Atlantic ASW exercise showed good results, and the SQR-14/15 Towed Array Sonar System (TASS) became operational, with four TASS units installed on DE-1040 class ships.

For noisy targets, the passive techniques were more attractive than the active techniques in that these systems were easier to operate and did not reveal the searcher direction to the submarine target.

In 1989, J. Richard Hill wrote the following:

> Surface ship active sonars have made little perceptible progress over the past 5 years. Partly this was due to the confidence with which passive means were being viewed 10 to 15 years ago, when active sonar tended to take a back seat.

Fifteen years from 1989 would place the year at which active sonar "tended to take a back seat" as 1974, which is consistent with the NUSC observations. Hill goes on to say the following:

> Clearly, the 1980's were the decade of the passive, particularly on the Western side; not only were targets helpful noise emitters, but equipments and processing were making rapid advances, the self-noise of platforms was being satisfactorily reduced, and the command and coordination of ASW assets were making considerable strides.[1]

Chapter 8 — Equipment Operation and Tactical Employment

After about 1974, it was commonly observed that SQS-26 ships made little or no use of their active capabilities. During the early 1980's, an experienced sonar chief petty officer who had served on an SQS-26 ship commented to me that he knew little about active sonar operation because his ship had never employed the SQS-26 in the active mode.

The effectiveness of passive sonar was to come to an end around 1989, with the development by the Soviets of quiet submarines — in part due to the efforts of John Walker's spy ring, which operated from inside the U.S. Navy from 1967 to 1985. Admiral James D. Watkins, former CNO, credited Walker with having given the Soviets the information that they needed to improve their submarine construction technology to compete more effectively with U.S. technology.

Defector Vitaly Yurchenko, a former KGB agent, said Walker gave them the ability to read over one million military messages over the years.[2] In addition, Walker provided the Soviets with actual copies of defense plans, logistics information, weapons characteristics, and tactical publications.

Although active sonar again became of interest after 1989, the Cold War ended in 1991. From that point forward, the interest in training operators to use active sonar never reached the levels of the 1970 to 1974 period.

CHAPTER 9
FLEET PERFORMANCE

IMPORTANCE OF INFORMATION RELATED TO FLEET PERFORMANCE

Developmental testing is performed in a sheltered environment in comparison to what is encountered in Fleet operations, whether they be free-play exercises against U.S. submarines or Cold War encounters against unfriendly submarines. Both problems and opportunities are encountered in free play that could not have been envisioned in developmental testing. As a consequence, it is essential for those engaged in system development to obtain information on the performance of new equipment that has been turned over to the Fleet. Of special interest in SQS-26 operations was any performance related to the use of convergence zone and bottom bounce modes — neither of which had been encountered by the Fleet in predecessor sonar equipments.

NUSL obtained the information on free-play performance from firsthand observations of NUSL ship riders or from messages and reports received at NUSL from Fleet units operating the SQS-26 equipment without NUSL assistance. The observations of experienced NUSL personnel were extremely valuable, providing a greater depth of information than could be expected from Fleet personnel. On the other hand, observations made in the absence of NUSL experts provided a better measure of what the ship could do during a more typical operational situation when assistance from NUSL would not be available.

The results of several types of Fleet experience with the systems were of interest:

- NUSL-prescribed Fleet training at the special locations off Guantanamo Bay, San Diego, and Pearl Harbor.
- Contacts of opportunity on surface traffic. The target reflectivity of surface ships is similar enough to that of submarines to provide a good indication of general SQS-26 sonar capability against a submarine about which nothing is known regarding range, bearing, speed, and course (a necessary condition is that ship personnel must be able to set up and operate the equipment).

Obtaining unalerted detections on surface ships and maneuvering to hold contact provided valuable practice for the ship's ASW team, as well as a good indication of SQS-26 effectiveness.

- Contacts of opportunity gained against random encounters with U.S. Navy, Soviet, or Allied submarines.
- Contacts during special surveillance operations set up to gain intelligence on Soviet submarine locations and movements.
- Contacts against submarines participating in free-play Fleet exercises sponsored by the U.S. Navy or Allied organizations.

A chronological sampling of this activity is provided next. Only those examples illustrating particular capabilities are presented.

SEPTEMBER 1965: OBSERVATION OF SQS-26 PERFORMANCE IN A FLEET EXERCISE

In September 1965, NUSL's Richard Chapman and Albert Silverio rode on USS *McCloy* (DE-1038) during an 8-day NATO convoy protection exercise that was opposed by six submarines. Conducted in the Atlantic Ocean off the northeastern coast of the United States, this event would provide the first opportunity for NUSL engineers to observe free-play submarine detections by an SQS-26.

McCloy was commissioned in 1963 as one of only two destroyer escorts in the 1037 class, both of which had received the first two production SQS-26 systems built by GE (the only SQS-26 systems with no suffix designator). These systems were similar to the SQS-26 (XN-2) as it existed prior to any improvements.

Environmental conditions were suitable only for surface duct operation. Using the surface duct ODT mode, *McCloy* obtained four free-play, completely unalerted detections at ranges between 10 and 22 kiloyards, corresponding to theoretical expectations. *McCloy* was involved in successful follow-up attacks on three of the four detections.

The initial SQS-26 production system had passed its first documented free-play test, at least for surface duct coverage, despite the lack of improvements that were later retrofitted into the SQS-26 and the similar SQS-26 (AX) systems.

JULY 1966: FREE-PLAY SUCCESS WITH THE CONVERGENCE ZONE MODE

In July 1966, USS *Brooke* (DEG-1), with an SQS-26 (AX), made two convergence contacts at 30 miles in an ASW exercise off San Diego. This was the first documented convergence zone detection in a free-play exercise. In January 1967, another free-play convergence zone contact was made by *Brooke* at 26 miles in the same exercise area.

These detections were surprising because *Brooke* had one of the first 12 SQS-26 (AX) production models. To meet shipbuilding schedules, these systems had been manufactured without the benefit of sea test results from the XN-2 experimental system and were notoriously difficult for the Fleet to maintain and operate. For one example, the AX employed the paper recorder system that presented an echo history of only the last ping. Earlier in 1966, OPTEVFOR had done an AX operational appraisal recommending that the system just be used for surface duct search, with NUSL in complete agreement. The AX systems were later extensively retrofitted and reclassified as AXRs.

After receiving a report of *Brooke's* favorable experience with the convergence zone mode, I visited the ship in Long Beach, California, and found an unusually motivated group of sonar operators with a refreshing "no-problem" approach to both operating and maintaining the AX equipment (despite its shortcomings). They were excited about their demonstration of a long-range convergence detection capability, as well as about their ability to vector an aircraft out to the ship's convergence zone datum for a follow-up attack. Knowing that the accuracy of the analog range determination measurement with the paper recorder display was poor, I asked how they determined ranges accurately enough to put an aircraft on top of the target. The operators replied that a commercial timing counter had been "rigged up" to provide the required accuracy between the time of ping transmission and echo reception.

I left *Brooke* with the feeling that perhaps the anticipated SQS-26 maintenance and training problems were not so troublesome after all. However, this was an overly optimistic assessment that would not be borne out by more experience with the production installations. I later concluded that only about 1 crew in 10 could cope with SQS-26 operation and maintenance problems in the early SQS-26 production systems, although, in retrospect, the number was probably somewhat greater than that. In any event, while the *Brooke* had an exceptional crew, the rest of the Fleet required considerable assistance. I would be reminded of young Harvard instructor Henry Adams and his gloomy assessment of the students: Those whose minds were above average were, in his experience, "barely one in ten; nine minds in ten take polish passively, like a hard surface; only the tenth sensibly reacts."[1] This comment, however, seemed to ignore the positive impact of leadership, inspirational teaching, and other environmental influences on individual capabilities.

The disproportionate contributions to Fleet capabilities of a small number of ships, whatever that number might be, is consistent with the Pareto Principle, named after the 19th-century Italian mathematician, engineer, economist, and sociologist Vilfredo Pareto. This principle states that the significant items in a given group normally constitute a relatively small portion of the total items. As NUSL collected more statistics, the Pareto Principle seemed to be confirmed. In the opportunities to interview personnel from the especially productive ships, the well-above-average attitude and competence of the personnel, as I had observed on the *Brooke*, seemed to be in accord with the exercise results obtained.

Norman Augustine compiled statistics in both military and other fields that tended to quantitatively confirm the Pareto Principle. In his data sample of many disparate occupations, 50% of the output was consistently contributed by only 20% of the participants.[2] His measures of achievement included military air-to-air victories, staff actions of the Joint Chiefs, NFL rushing touchdowns, industrial patents, and papers contributed to journals.

MAY 1968: FEASIBILITY STUDIES OF CONVERGENCE ZONE APPLICATIONS IN THE MEDITERRANEAN SEA

Request from the Sixth Fleet for Assistance

In May 1968, a letter from the Commander, ASW Force Sixth Fleet (COMASWFORSIXTHFLT), requested that NUSL provide the technical expertise necessary to permit exploitation of the full capabilities of operational ASW sensors in the Mediterranean environment. This letter was followed by a visit to NUSL from Captain Fred J. Kelly of COMASWFORSIXTHFLT on 13 June 1968. During the visit, he explained how bad the conditions were for direct path sonar detection, especially in the warm months of the year when severe negative gradients caused downward refraction that limited detection ranges to a mile or less.

ASW was of particular concern at that time because the Soviet Fleet had been increasing its submarine presence in the Mediterranean Sea for several years, with the 1968 steady state level approaching eight Soviet submarines — up from one to two submarines in 1965. Norman Polmar tells us that in 1956 the Soviets (all ship types included) spent only 100 ship days in the Mediterranean Sea, whereas in 1965 their presence had grown to 5,600 ship days and by 1970 it was 17,400 ship days.[3]

Investigating the Feasibility of Using the Convergence Zone Path

It was concluded that the most profitable initial direction for NUSL assistance would be to investigate the operational utility of employing both the SQS-26 and SQS-23 sonars to exploit the convergence zone path in the Mediterranean. While convergence zone ranging was expected to be normally beyond the capabilities of the SQS-23, the short range of convergence zone formation in the Mediterranean opened up the possibility of using that system for such a search. The initial deployment of SQS-26 systems in the Mediterranean involved only BX models.

NUSL's Proposed Program

As a result of the conference with Captain Kelly, NUSL agreed to undertake convergence zone investigations that would initially be of a theoretical nature but would eventually lead to experimental work. The objectives of the convergence investigations were to determine the following:

- Echo-ranging performance potential for both the SQS-26 and SQS-23 systems,
- Optimum sonar operating procedures,

- Optimum ship employment tactics, and
- Equipment and training deficiencies that might require corrective action.

Studies of Convergence Zone Path Possibilities in Mediterranean

During the months that followed, NUSL systems engineering studies (in my group) were directed especially at the deep-sound velocity structure of the Mediterranean Sea,[4] its near-surface temperature variation with time and location,[5] and the possibility of using the SQS-23 for convergence zone echo-ranging. Emphasis was placed on the SQS-23 in the studies because there was no history of how this system would perform during convergence zone search and also because of its predominance in the Mediterranean force levels at that time. For the SQS-26, considerable experience had already been obtained during convergence zone testing in the Atlantic and Pacific, although the short convergence zone ranges at certain times of the year in the Mediterranean Sea opened up the possibility of multiple convergence zone echo-ranging. The depths required for convergence zone operation were examined, along with the expected convergence zone ranges as a function of location and season. Also examined by NUSL were charts of plankton density as a function of location to provide an estimate of the seriousness of biological reverberation. With this information, possible tactics were considered for using the convergence zone geometry to conduct a barrier patrol, perform a broad area sweep, or screen a battle group.

Figure 15 shows typical SQS-26 direct path and convergence zone coverage in the Mediterranean versus month for a submarine at shallow operating depth.[6] In the 5 warm months of the year (May through September), no surface ducting is commonly present and direct path ranges typically are environmentally limited to about 1.5 kiloyards. *For the older sonars capable of using only the direct path, 1.5 kiloyards represents the maximum submarine detection capability in these 5 warm months.* This detection capability was worrying Captain Kelly when he came to request NUSL's assistance.

As seen in figure 15, the NUSL study showed that in the warm months a system such as the SQS-26 could reliably exploit convergence zone paths to obtain maximum range coverage from 35 to 50 kiloyards.

Chapter 9 — Fleet Performance

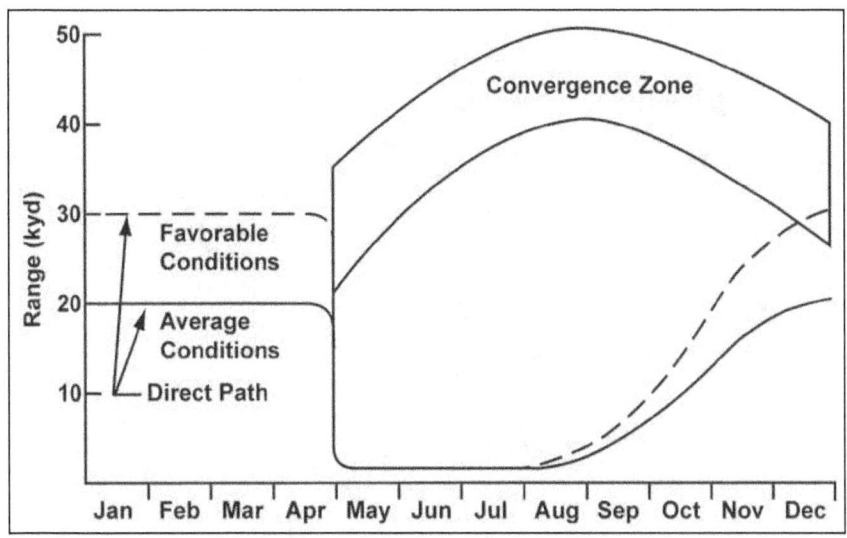

Note: The very short detection ranges via the direct path in the warm months of the year were a major problem for predecessor sonars.

Figure 15. Direct Path and Convergence Zone Detection Coverage of Shallow-Water Submarines by the SQS-26 in the Mediterranean Sea

Moreover, it was found that convergence zone ranges were not sensitive to submarine depth.

From late fall through winter and into early spring marks a transition period where the thermal structure gradually changes from summer conditions to winter and back again. The corresponding propagation conditions are also in transition from direct path only to convergence zone, and then back again to direct path. Whether the conditions represent deep ducting or convergence zone propagation, the SQS-26 ranges will still typically run from a minimum of 20 to a maximum of 50 kiloyards.

Figure 16 shows the ray paths for these transition months in the Mediterranean. Note that small changes in the upper 100 or 200 feet can change the picture from a convergence zone to a surface duct condition. In both cases, the propagation conditions are favorable out to over-the-horizon ranges and can be fully exploited over this region with the SQS-26.[7]

Chapter 9 — Fleet Performance

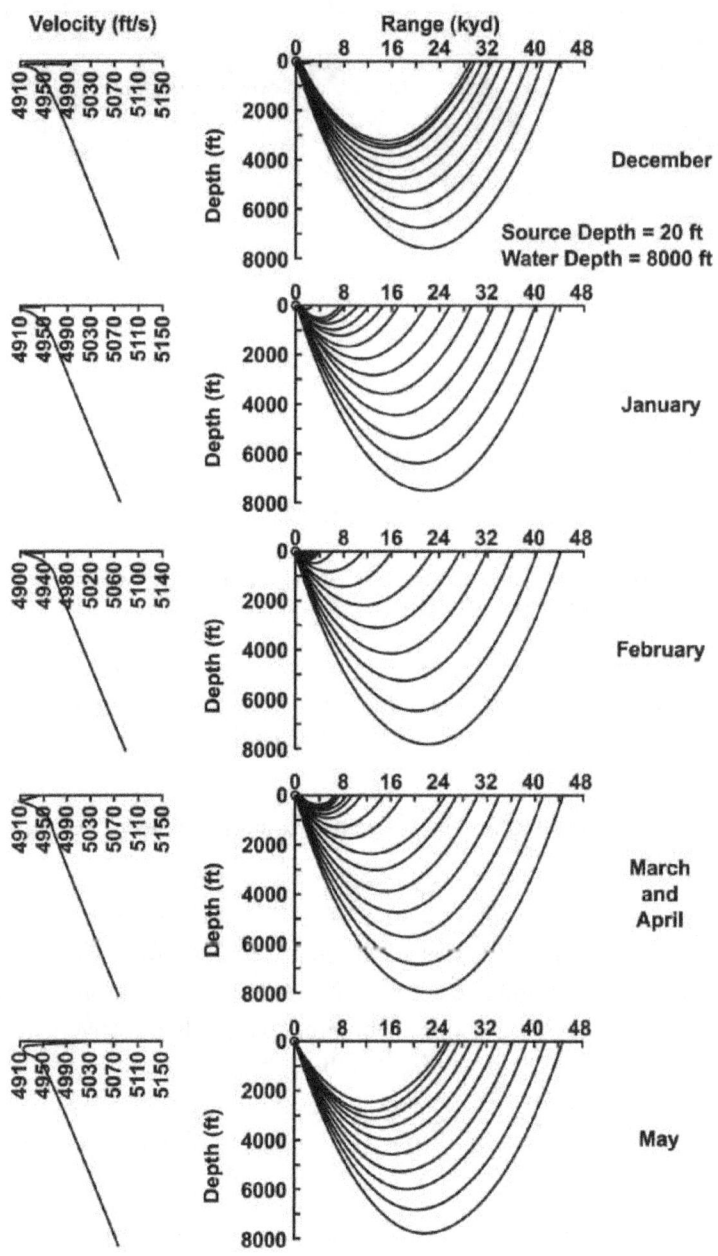

Note: Small changes in the thermal conditions in the upper few hundred feet produce the change from continuous to zonal coverage. In any event, the maximum ranges are long and exploitable by the SQS-26.

Figure 16. Ray Paths During the Transition Months from Convergence Zone to Direct Path and Back Again

Chapter 9 — Fleet Performance

Presentation to ASWFORSIXTHFLEET

As these studies were completed, by August it was decided that enough material had been generated to warrant a presentation of the results to ASWFORSIXTHFLT. Accordingly, on 11 September 1968, NUSL representatives gave a status presentation in Naples, Italy, to Vice Admiral E. C. Outlaw (COMASWFORSIXTHFLT) and his staff.

Controlled Convergence Zone Testing

As a consequence of the presentation, 2 days of controlled convergence zone testing was scheduled to begin 31 October 1968. A submarine target and two ships — one with an SQS-26 (BX) and the other with an SQS-23 —would be employed. The instrumented submarine target would permit measurement of propagation loss.

NUSL's Bernard Cole directed the SQS-26 (BX) testing and Harold J. Doebler handled the SQS-23 testing. Strong first-zone echoes were obtained at all aspects with the SQS-26, but results with the SQS-23 were marginal at nonbeam aspects. Second-zone echo-ranging attempted with the SQS-26 was unsuccessful. Echo-ranging at night was adversely affected by biological reverberation, which was not unexpected based on previous North Atlantic and Gulf of Mexico experimentation.

Unexpectedly High Attenuation

Analysis of the convergence propagation loss measurements revealed higher than expected losses in both the first and second zones. At that time, NUSL was using William Thorp's attenuation as determined on the sound channel axis in the North Atlantic. It had previously been noted that convergence zone propagation loss measurements indicated higher attenuation in the Atlantic than in the Pacific. Now, the attenuation in Mediterranean appeared to be still higher, as if there were some deleterious effect on attenuation due to the change in longitude.

Juergen Keil in an analysis of 385 convergence zone contacts made by SQS-26 ships between June 1970 and June 1973 astutely observed[1] that the Mediterranean zone widths over which targets were held were markedly narrower than those calculated with the Thorp model. He found that the two-way sonar equation seemed for some reason to be some 8 dB in error in the Mediterranean. Since the equation included equipment figure of merit uncertainties, Keil postulated that the 8 dB might be accounted for by equipment degradations due to the ships in the Mediterranean being further away from U.S. major support centers. It would be 1988 before Robert Mellen's quantitative calculations confirmed that convergence zone path attenuation in the Mediterranean would be markedly higher than that predicted by the Thorp model. Mellen's refined model predicted the error in the Mediterranean (resulting from use of the Thorp relation) to be actually 12 dB rather than Keil's inferred 8 dB from looking at measured versus calculated zone widths.

Mellen's 1988 model indicated that the Mediterranean attenuation per kiloyard over the convergence zone path would be approximately double that over the North Pacific convergence zone path, with the reason for these differences in attenuation due to chemical differences in the sea water found in each location. Mellen also determined that attenuation in the deep sound channel axis as measured by Thorp was significantly different from that expected over the convergence zone path because of the change in water chemistry (and therefore attenuation) with depth.[8]

Both Mellen and Keil had indicated the existence of a major error in the Thorp model for predicting Mediterranean performance. Considering the many uncertainties that Keil had to deal with in the calculation, his results compared with Mellen's later calculations were not at all unreasonable.

Nearly a quarter of a century passed between the time when the SQS-26 attenuation differences between the Pacific and the Atlantic were first observed and the time when Mellen published a model of convergence zone attenuation that accounted for such worldwide differences. For years, no one understood what was happening until Mellen, with contributions from many other investigators, provided the answer. In the interim, everyone wondered why performance seemed to inexplicably vary with location.

Some had believed that the attenuation measurement was so difficult to make that quantitative observations of the differences being addressed were close enough to the error of measurement that the observed differences were not significant. It is interesting to note, however, that some unexplained echo-ranging performance change with location was consistently being seen over 25 years. The Mellen model, which depended on location-dependent chemical changes in sea water, finally quantitatively vindicated the attenuation observations, with NUSL echo-ranging experiences helping to validate the reasonableness of the model.

Impact of High Attenuation

The impact of the unexpectedly higher attenuation found in these and later experiments in the Mediterranean was as follows. For the SQS-26, the generally shorter geometrical ranges to the first zone in the Mediterranean (as compared to those encountered in the Atlantic and Pacific) largely offset the higher attenuation. Performance at the second zone, however, was generally not reliable and was well below original expectations. Performance with the SQS-23 was marginal, even in the first zone. In retrospect, these disappointments in performance were clearly attributable to the inadequacy of the Thorp model for Mediterranean predictions.

Briefing on the Outcome of Tests

On 5 November 1968, NUSL briefed ASW Group One (ASWGRU-ONE) in Naples on the results of the controlled tests. Search tactics that would exploit the demonstrated convergence zone detection capability were also suggested at that time. A trial of these tactics was conducted during an ASW search operation that had already been scheduled for 7 November 1968 in the Tyrrhenian Sea. Within the time span of only 1 month, the results of preliminary studies had been presented to COMASWFORSIXTHFLT, follow-on controlled experimentation had been completed, and experimentation with operational use of the SQS-26 in the convergence zone mode had begun.

NOVEMBER 1968: CONVERGENCE ZONE SWEEP OF THE TYRRHENIAN SEA WITH TWO SHIPS

There would be two SQS-26 (BX) ships available for the ASW sweep in the Tyrrhenian Sea: USS *Voge* (DE-1047) and USS *Koelsch* (DE-1049). Captain Marty Zenni, the surface ship squadron commander, was on *Voge* (as was I), and Bernard Cole was on *Koelsch*.

Chapter 9 — Fleet Performance

Because Captain Zenni had heard that naval personnel did not always take advantage of civilian expertise at sea, he appointed me as chairman of a committee that would determine how to best conduct the search operation. A procedure was worked out that employed the two SQS-26 ships in a line-abreast formation, with two other ships of the squadron (*Dealey* class escorts) positioned out ahead in the SQS-26 convergence zone area. An ASW carrier also participated in the sweep operation.

The *Dealey* class escorts would search with their sonars passively and conduct a radar search of surface traffic to assist in the classification of any active sonar targets detected by the SQS-26 ships. They would also be available to be vectored toward any submerged targets detected by the SQS-26 systems on *Voge* and *Koelsch*. With the *Dealey* escorts positioned in the convergence zone coverage region about 20 miles ahead, they could provide a check on the SQS-26 detection capabilities. Although the two SQS-26 ships were initially positioned more than one convergence zone apart laterally to maximize the combined coverage, they were later moved into a first convergence zone, 20-mile spacing to ease station keeping and to permit monitoring of the SQS-26 sonar capability by mutual echo-ranging.

Visit from CVS Chief of Staff

In the early stages of the operation, the Chief of Staff on the CVS (an ASW aircraft carrier) visited *Voge*. The commanding officer brought him into the combat information center (CIC), where a radar picture showed the location of the other ships in the sweep formation. Just as they were entering the CIC, the radar operator reported loss of contact on the escorts in the forward sector. I announced that this would not be a problem because the two over-the-horizon *Dealey* escorts were being tracked at 20 miles on sonar, which was probably the first observation of production sonar equipment outperforming radar against a surface target. At the time, all the ships were moving at a speed of about 10 knots, which permitted *Voge* a good "opening" Doppler indication on the ships in the zone with the CW Doppler detection waveform. Own ship motion is nullified in the system so that any significant target motion through the water, as was the situation with the ship in the convergence zone, shows up as Doppler — in this case 10 knots in the direction away from own ship.

Chapter 9 — Fleet Performance

Results of ASW Sweep

During the 4-day search period, the only submerged targets detected were sea mounts. However, because *Voge* and *Koelsch* were receiving consistent echoes from each other on the convergence zone path, a reasonably good confidence level existed regarding their detection capability over the combined two-ship sweep width of about 45 miles in the line-abreast formation. These surface ship target strengths were comparable to those of a submerged submarine. It was estimated that some 14,000 square miles (adjusted for equipment downtime, fueling, and sonar degradation during a day of gale force winds) had been searched during the operation.

The lessons learned from this experience would be applied in the months ahead to more successful demonstrations of ASW search operations. Arrangements were made for NUSL's Doebler to remain in Naples for a period of at least a year where he would provide expertise to the staff for exploiting convergence zone paths in ASW operations.

NUSL would brief ships leaving the United States for the Mediterranean on the results of SQS-26 experiences. Further instruction, provided by Doebler and other NUSL engineers after the ships arrived in the Mediterranean, described the environmental cycles of the various sonar paths, tactics for exploiting the paths, and results of past operations against submarines. Technical problems encountered by the ships were referred back to NUSL for appropriate action.

Doebler was provided with assistance from NUSL in various forms. When he became temporarily ill, I traveled to Naples to assist with his schedule, which included a trip to Cannes to brief the carrier group on Mediterranean ASW techniques. Figure 17 is a photograph of my visit in 1968 to USS *Little Rock* (CLG-4), the Sixth Fleet flagship.

Why Sea Mounts Can Look Like Submarines

Twenty years later, shortly after I had retired from government service, I conducted a study for NUSC on the classification of sea mounts in deep water, based on the 1968 experience in the Tyrrhenian Sea sweep.[9] From this 1988 investigation of sea mount contacts made during that earlier sweep in the Tyrrhenian, I discovered that a special

Chapter 9 — Fleet Performance

Figure 17. Thaddeus Bell (the Author) During a Visit to USS Little Rock (CLG-4) in the Mediterranean Sea

acoustic condition was required for a sea mount to present a submarine-like target — an intersection between a sea mount and a sound field caustic (or "hot spot") at a particular depth under an upward refraction condition. The hot spot was similar to that formed near the sea surface in the convergence zone, and, when concentrated, it presented a short enough echo in the range dimension from the sea mount to resemble a submarine echo. Otherwise, the reflection from the sea mount would be extended and blend in with the bottom reverberation, which explained why only a few of the many charted sea mounts in the path of the sweep presented submarine-type echoes.

Most sea mounts did not extend far enough upward from the bottom to intersect with a caustic. However, for those that did, when the caustic moved over the sea mount surface as the ship approached, the target range closure would be amplified such that a closing target range rate would appear to be produced by the stationary mount.

MARCH 1969: CARRIER SCREENING EXERCISE OFF SAN DIEGO

In March 1969, Richard Chapman and I boarded USS *Bronstein* (DE-1037) to participate in the hunter-killer ASW exercise (HUKASWEX) 3-69, a Pacific event of several days duration in which experiments were conducted with the SQS-26 for carrier protection. By that time, *Bronstein's* SQS-26 sonar equipment had been upgraded to an SQS-26 (AXR).

Prior to departure from Long Beach, I briefed the CVS staff on exercise tactics that placed *Bronstein* on the flank of the carrier, where it would use the SQS-26 (AXR) to provide a submarine detection screen over a 240° arc in the threat direction. Although the required size of the arc was larger than optimum for SQS-26 coverage capabilities, there was only one SQS-26 ship available.

During the exercise, 4 detections out of 10 opportunities in the search sector were obtained, which was quite satisfactory considering the following problems. First of all, the size of the coverage arc went beyond the equipment design objectives, requiring somewhat awkward manual time-sharing to cover the required 240°. Next, the operators had no previous experience in this type of operation. Finally, the attempted 28-mile convergence zone range coverage was pushing the expected 30-mile convergence zone limit for steel dome ships. In any event, *Bronstein* demonstrated a single-ship submarine detection coverage perimeter of some 28 miles, independent of submarine depth, which was unprecedented.

During one of the detections in the sonar control room, I noticed that the echo was showing a nearly zero range rate and appeared to be a zonal hot spot artifact because it was going along with the zone rather than changing range. Post-exercise analysis, however, showed that it was indeed a real submarine contact, where the submarine happened to be running nearly parallel on *Bronstein's* flank, closing range very slowly. In previous experience with controlled tests, the submarine was always placed ahead of the ship's path, traveling on a nearly opposite course, which produced a substantial range rate. These earlier scenarios had resulted in "negative training," as controlled tests sometimes do. The *Bronstein* exercise, on the other hand, taught the lesson that, in real life situations, the submarine will not always show a substantial closing range rate but, in fact, may be on courses and at speeds such that it will linger for a considerable time in the covered zone. For the detection opportunities in the convergence zone at 28 miles, the submarine (unaware that it was being tracked) was focused on its main objective — the high-value carrier unit that was being protected. The submarine's problem was trying to decide on a best approach course and that did not necessarily mean heading in the direction of the *Bronstein* escort.

On another convergence zone sonar contact that looked genuine, as indeed it was, the commanding officer of the carrier attempted to vector one of the helicopters from the CVS out to the contact. All was going well until the helicopter (reporting low fuel) was forced to return to the carrier. This very important event would have been the *first* helicopter "attack" on a convergence zone contact provided by a surface ship.

On a third contact, information was obtained that indicated a serious design flaw in the SQS-26. The contact was made initially on a closing-range-rate submarine target, which about half way through the zone started to change course and finally open range. It was discovered later that the target was not aware of being in the coverage zone. Only by coincidence was the opening maneuver made by the target as it was being tracked in the zone. Its speed also began to increase, as a submarine will often do when it runs to another location to create a baseline for a new acoustic observation. As the speed increased to above 20 knots, contact was lost, even though the submarine was being kept within the zone. At the time, I did not understand why contact was lost.

When I related the experience to William Downes back at NUSL, he asked me to think about why the submarine faded out above 20 knots of opening Doppler. He reminded me that in the system specifications there was only enough bandwidth provided to allow for 20 knots of down Doppler on the rationale that a destroyer's main interest would be in detecting closing targets — not those going away at more than 20 knots. There was no provision for a scenario that involved tracking a target that had turned away at a speed of more than 20 knots. Downes immediately acted to develop a field change to all SQS-26 systems that would allow for an appropriate opening Doppler that extended well above 20 knots.

JULY 1969: SUCCESSFUL SWEEP OPERATION IN THE CONVERGENCE ZONE

In June 1969, Commander William A. Myers III was assigned to the Sixth Fleet in the Mediterranean as Commander of Destroyer Division 262.[10] In July, he embarked on *McCloy* in Naples to conduct a series of planned measurements, as well as free-play exercises.

NUSL's Bernard Cole joined *McCloy,* which was equipped with an SQS-26 (AXR), for a July ASW sweep of the Ionian Sea. The convergence zone sweep width at that time of the year was about 40 miles. Table 2 summarizes the high points of this operation.

Table 2. July 1969 Sweep Operation in the Convergence Zone

Date/Time	Event
071200	Start convergence zone sweep of the Ionian Sea.
080742	*McCloy* gains convergence zone contact at 41 kiloyards after a search of nearly 20 hours.
080817	*McCloy* vectors a VP (ASW land-based patrol aircraft) to datum within 1 hour after sonar contact is made. VP sights bubbles, obtains passive signal with sonobuoys, and makes magnetic anomaly detector (MAD) contact.
080829	VP starts to echo-range with the sonobuoy. Submarine goes into high-speed evasion, and contact is lost. *McCloy* is later diverted to a search and rescue (SAR) operation.
090137	About 17 hours after contact is lost, VP sights submarine on surface 18 miles southwest (SW) of datum, heading SW. Identifies submarine as Soviet *Foxtrot* class. Submarine dives and VP contact is lost again.
090915	*McCloy* released from SAR. Searches probability area based on submarine course and speed at last VP sighting.
100330	*McCloy* regains convergence zone contact about 6 hours after search is resumed. Intercepts radar transmissions and vectors VP into MAD contact range.
101749	About 14 hours after contact is regained, contact is lost.

Chapter 9 — Fleet Performance

JULY 1969: CONVERGENCE ZONE EXERCISE VECTORING A POUNCER INTO ATTACK RANGE IN THE TYRRHENIAN SEA

Later in July 1969, *McCloy* participated in an exercise with a U.S. conventional submarine that was constrained to operate within a 1° square in the Tyrrhenian Sea, evading detection if possible. In addition to *McCloy,* an SQS-23 ship — USS *Barney* (DDG-8) — would act as a pouncer and a VP aircraft would provide general ASW support. No NUSL engineers were aboard *McCloy*. Results of this sweep are summarized as follows:

- At 0023, VP obtains the first detection (a radar indication from the submarine's periscope).
- Fifteen minutes later, the submarine spots the VP and dives to 120 feet. The VP obtains MAD plus active sonar contact.
- Twenty-three minutes later, the VP loses all contact.
- Two hours later, *McCloy* gains an active convergence zone contact.
- One hour later, *McCloy* vectors VP into MAD range.
- Ten minutes later, VP sights periscope.
- Two hours later, *McCloy* vectors *Barney* into SQS-23 direct path active sonar range at 0.6 kiloyards.
- *Barney* makes successful ASW attack.

This appears to be the first exercise in which an SQS-26 made a convergence zone contact that was followed up by the vectoring of an SQS-23 surface ship into direct path attack range.

AUGUST 1970: BOTTOM BOUNCE PERFORMANCE TESTS IN THE MEDITERRANEAN SEA

Controlled Testing of Bottom Bounce Mode

In August 1970, John Hanrahan of NUSC conducted the first controlled bottom bounce testing in bottom conditions that were typical of the Mediterranean Sea. USS *Glover* (AGDE-1) with an SQS-26 (AXR) made 10 closing runs on a conventional U.S. submarine operating at 5 knots and a 60° aspect at ranges out to 25 kiloyards. *Glover*

Chapter 9 — Fleet Performance

maintained a speed of 10 knots. Over-the-horizon bottom path detections were made on 9 of the 10 runs.

Free-Play Bottom Bounce Tests

The controlled testing was followed by a free-play exercise conducted by the Destroyer Development Group against a U.S. conventional submarine that had only the one restriction of operating in a given lateral sector. Because *Glover* was to search this sector, simulating a convoy protection scenario, it adopted speeds and search tactics that NUSC would not have recommended (i.e., ATP-1 (allied tactical publication 1) evasive steering and speeds of 16 knots). For the steel domes of that era, NUSC did not recommend speeds that were higher than 12 knots.

Despite this, *Glover* obtained bottom bounce detections on 6 of the 10 runs. Moreover, whenever the echo-ranging *Glover* maneuvered in a manner consistent with maintaining contact, tracking was demonstrated into the weapons range below 10 kiloyards, *marking the first successful free-play demonstration of SQS-26 bottom path detections in the Mediterranean.*

Extrapolating this experience to the rest of the deep-water (greater than 1000 fathoms) locations in the Mediterranean through the use of all available data on bottom characteristics, satisfactory performance was predicted for the bottom bounce path in the great majority of Mediterranean deep-water locations (given reasonable search tactics). One reason that little had been attempted previously regarding use of the bottom bounce mode in the Mediterranean was that there was a limited understanding of bottom loss variability in that area. It was not until May 1971 that NUSC's Eugene Podeszwa formally published the first bottom province chart for the Mediterranean from his analysis of the MGS data.

APRIL 1971: FREE-PLAY CONVERGENCE ZONE EXPERIENCE ON *HORNE* FOR HOLDEX 2-71 IN THE PACIFIC OCEAN

USS *Horne* (DLG-30) turned in one of the more interesting reports of free play using the convergence path. From 30 April to 5 May 1971, *Horne* employed its SQS-26 (BX) for HOLDEX 2-71 in the Pacific. During the exercise, nine convergence zone contacts were generated and held for up to 2 hours. Shipboard tactics proposed by NUSC to maintain a contact in the convergence zone were tested, and techniques for interplatform coordinated tracking of the submarine were formulated. Results of this exercise led to the following observations by *Horne:*

- *Experience in gaining contact indicated the importance of rapid action.* The sonar operators quickly grasped the fact that information on the initial video return must be disseminated promptly if contact were to be held. The ship must be immediately maneuvered so as to hold contact (in accordance with the tactics in naval warfare interim publication (NWIP) 1-4(B) developed by NUSC).

- *The importance of coordinated operations became apparent.* In one case, as *Horne* detected the submarine conducting an "evasion-unlimited" transit, it also vectored out VS aircraft followed by surface units to an estimated intercept point. These units held contact on the submarine for an extended period.

- *An important advantage of convergence zone detection capabilities was confirmed.* The submarine's heretofore advantage in blue water ASW operations had been its unique ability to go beneath the layer and then, due to inherent environmental and physical factors, evade the close-in surface units. This tactic of the submarine, however, appeared to be no longer valid against a ship operating in the convergence zone mode.

AUGUST 1971: SHALLOW-WATER TESTING ON THE TUNISIAN SHELF

From 29 August to 1 September 1971, NUSC's Bernard Cole conducted a series of controlled shallow-water, echo-ranging tests with *Glover's* SQS-26 (AXR). These tests — performed in cooperation with the Supreme Allied Command, Atlantic (SACLANT) Undersea Research Center — were conducted south of the island of Lampedusa on the Tunisian shelf at a water depth of about 200 feet.

The target submarine was instrumented to permit propagation loss measurements. A median active detection range of 15 kiloyards was obtained during 13 runs on the 3-knot target over a variety of aspects from beam to bow. The minimum range was 12 kiloyards and the maximum range was truncated at 23 kiloyards,[11] which was consistent with Fleet results on shallow-water submarine detections analyzed by NUSC during the period from 1971 through 1974. The previous shallow-water Fleet detection ranges reported were from 7.9 to 44.9 kiloyards, with a median of 14.8 kiloyards.

SEPTEMBER 1971: VALIDATION OF THE NUSC ACOUSTIC PROVINCE CHART

As a result of the successful August 1970 bottom bounce tests in the Mediterranean Sea, COMASWFORSIXTHFLT requested that NUSC validate its acoustic province chart over as much of the central and eastern Mediterranean as possible during 10 days at sea during September 1971. The eastern region was chosen because of its strategic value, the paucity of active sonar data there, and indications from its physiographic bottom core and MGS data that conditions would be favorable for bottom bounce echo-ranging. The resulting test program (known as "Med 71")[12] was planned and directed by NUSC's John Hanrahan.

Two ships were involved, USS *Jallao* (SS-368) and *Glover*. The tests were conducted along a 1300-mile track that extended from Malta to a point north of Egypt and then to a region off the western coast of Cyprus. Measurements were made of propagation loss via the bottom path as a function of depression angle at each of 17 stations along the track. In addition, measurements were made of echo-ranging reception, bottom backscattering, wind speed, and thermal profiles.

The tests were a reaffirmation of previous assertions that the SQS-26 possessed a bottom bounce search capability in all the NUSC bottom loss province classifications up through 4 and possessed a tracking capability up through 5 in the Mediterranean Sea. The bottom loss versus angle relation was found to be much more variable in the eastern Mediterranean than it was in the western Mediterranean. Tactically, this meant that a submarine detection in the eastern Mediterranean should be exploited by holding the target at a constant range while another platform is vectored out to deliver the follow-up attack. In the western Mediterranean, however, it would be usually feasible to track the target continuously to within own ship weapon range.

Chapter 9 — Fleet Performance

SEPTEMBER 1971: SEMIFREE-PLAY CONVERGENCE ZONE EXERCISE IN THE MEDITERRANEAN SEA

A structured convergence zone penetration exercise was run by Commander, Destroyer Development Group (COMDESDEVGRU), in September 1971 in the Mediterranean Sea. This was described as a "semifree-play" exercise because of the existence of certain artificialities that would not be included in a typical free-play exercise. The conventional submarine was initially provided with significant knowledge of the surface ship range, search plan, and intended movement. The surface ship, *Glover,* expected a penetration attempt within a time window of an hour or so. *Glover* knew the general threat sector, although the exact target bearing was unknown.

The submarine was allowed to assume any course, speed, and depth after it believed detection had occurred, but took such action on only 3 of the 13 runs. Apparently, it had difficulty in determining when detection had occurred. When *Glover* made detection (median penetration aspect was 20°), it attempted to hold contact for an hour, after which period the run was terminated. Detection was made on 11 of 13 runs, with tracking times after detection varying from 14 to 78 minutes for an average of 49 minutes. Detection range averaged 45 kiloyards and tracking zone width was 3.7 kiloyards.

OCTOBER 1971: FREE-PLAY CONVERGENCE ZONE DETECTION AND ATTACK OPERATIONS WITH TWO SHIPS

In October 1971, the first free-play results in coordinated convergence zone operations with two SQS-26 ships were obtained in the Mediterranean for an exercise called "CZ Free-Play Noose." Convergence zone contact was established during each of two sweeps and was maintained while an assist ship was vectored in for attack. Contacts were considered significant in that the bow aspects presented by the U.S. conventional submarine did not prevent detection.

During the second sweep, USS *Connole* (DE-1056), with an SQS-26 (CX), gained convergence zone contact at 40.3 kiloyards. An attack aircraft was vectored on top 8 minutes later, confirming a snorkel that correlated with *Connole's* datum. USS *Daniels* (CG-27), with an SQS-26 (AXR), gained a convergence zone contact 10 minutes later at

41 kiloyards, which correlated with *Connole's* datum. Upon gaining convergence zone contact, *Daniels* paralleled the target's estimated course and speed to maintain contact in the annulus, where it was held continuously for 1 hour and 42 minutes. *Daniels* then vectored *Connole* toward contact with *Connole* passive. After *Connole* obtained passive contact, it went active at about 7 kiloyards and gained active contact at 6.4 kiloyards. *Connole* made an ASROC attack, dropping two practice depth charges (PDCs). Communications were established and the submarine surfaced at datum. It was later learned that the submarine had been unable to determine whether or not it had been detected in the convergence zone annulus.

DECEMBER 1971: CONVERGENCE ZONE CONTACT DURING A RANDOM ENCOUNTER WITH A U.S. SUBMARINE

During a routine convergence zone search operation in the Mediterranean in December 1971, USS *Belknap* (DLG-26), with an SQS-26 (AX), detected a U.S. nuclear submarine at communication depth, which was the first documented convergence zone detection to be obtained during a random encounter with a U.S. submarine. Submarine identity was verified by voice communications. Contact was held from 31 to 34 kiloyards, but operational commitments precluded maneuvering to hold the contact in the annulus

Although it had been generally accepted that the AX model did not typically have a convergence zone detection capability, *Belknap* apparently had not been informed of this limitation.

JANUARY 1972: CONTINUED SUCCESS BY *BELKNAP*

In January 1972, *Belknap,* still equipped with the SQS-26 (AX), continued its surprisingly effective performance in the convergence zone search mode in the Mediterranean. Four convergence zone contacts were made on conventional exercise submarines, with one exercise contact that demonstrated helo-surface coordinated operations against USS *Corporal. Belknap* held contact between 30.5 and 34.7 kiloyards as it vectored a light airborne multipurpose system (LAMPS) helo on top of the submarine, where the helo successfully gained MAD/sonobuoy contact.

Chapter 9 — Fleet Performance

As the opportunity presented itself, *Belknap* made 22 more convergence zone contacts on surface ships, many of which were gained under difficult own speed or wind-speed conditions. Contacts were made at own ship speeds up to 27 knots and at wind speeds up to 26 knots.

From January to April 1972, a total of 32 convergence zone detections against surface ships and submarines in the Mediterranean were reported and analyzed by NUSC. These convergence zone detections were reported by two ships: 26 by *Belknap* and 6 by USS *W. S. Sims* (DE-1059).[13]

APRIL 1972: *SIMS* CONVERGENCE ZONE PERFORMANCE WITH A SOVIET SUBMARINE AS THE TARGET

Sims, the other standout performer (in addition to *Belknap*) from January to April 1972, turned in a fascinating account of sonar operations against Soviet surface and submarine targets. *Sims* and USS *Pratt* (DLG-13), part of CTU 67.5.0, were tasked on 30 March 1972 to maintain surveillance of a Soviet *Foxtrot* class submarine in the company of a group of Soviet surface ships that were practicing ASW operations with the *Foxtrot* in the Gulf of Hammamet.

Foxtrot was sometimes on the surface and sometimes submerged. During an 8-day period, the Soviet submarine dove six times for a total of 38 submerged hours. *Sims* maintained almost continuous sonar contact during the submerged periods, making contact in the direct path at 7 kiloyards at one point. With *Pratt* maintaining contact at close range, *Sims* opened out to a convergence zone range of 30.8 kiloyards and then tracked the *Foxtrot* for 2 hours, at which point the *Foxtrot* surfaced.

On another occasion during this period, *Foxtrot* was on the surface with three U.S. Navy and three Soviet ships nearby. *Sims* opened to 30 kiloyards, from which range it was able to hold all surface ships and the submarine in the convergence zone window. After tracking the submarine for the next 8 hours, *Sims* was directed to proceed to another assignment.

Chapter 9 — Fleet Performance

During the foregoing operations, submarine contact was maintained despite (1) the *Foxtrot's* evasive maneuvers and generation of false contacts and (2) the Soviet surface ships' employment of "shouldering" tactics (i.e., "getting in the way of" a ship believed to be tracking their submarine). *Pratt's* shorter ranges meant that it operated closer to *Foxtrot*, where it deflected Soviet shouldering resources away from *Sims*.

APRIL 1972: *SIMS* RANDOM ENCOUNTERS WITH U.S. NUCLEAR SUBMARINES

On 9 April 1972, *Sims* was conducting Sixth Fleet operations in the Tyrrhenian Sea when it experienced a chance encounter at 24 kiloyards with a U.S. submarine during normal convergence zone search. Sonar operators had been issued no alert regarding the presence of the submarine in the area.

Upon reporting contact, *Sims* was directed by the task force commander not to prosecute. (The contact was subsequently identified by higher authority as a U.S. nuclear submarine on operations in the area.) As the range was allowed to open, the contact was tracked until it faded out at approximately 48 kiloyards. Subsequently, contact was regained in a very weak "second window" at 57 kiloyards, but then immediately faded. Total contact time was 2 hours and 15 minutes.

On 22 April 1972, *Sims* was again conducting a convergence zone search during a Sixth Fleet operation in the Ligurian Sea northwest of Corsica. Contact was gained at 40 kiloyards in the convergence zone window, with operators again having no previous knowledge of a submarine in the area.

Range was closed to 16 kiloyards by *Sims* on this new contact using the bottom bounce track mode. The contact was highly evasive, moving at speeds from 5 to 20 knots, with radical course changes. *Pratt* joined and gained contact at 16 kiloyards in surface duct mode. After the contact was positively identified as a known U.S. nuclear submarine on patrol in the area, *Sims* and *Pratt* were ordered to break off prosecution. As range was allowed to open, contact was tracked out to 38 kiloyards in the convergence zone mode before the submarine echo faded.

The contact had made clear, sharp, well-defined presentations on all displays throughout the prosecution.[14]

Chapter 9 — Fleet Performance

JULY 1972: DECLINE IN REPORTED CONVERGENCE ZONE CONTACTS

After mid-1972, the incidence of convergence zone contacts reported by the Fleet began to drop off. There were a number of reasons for the decline:

- First of all, the tempo of operations decreased because of various budgetary pressures and restrictions in fuel expenditure.

- Next, increasing inflation was raising the cost of military operations. The consumer price index that was 3.4% at the start of 1972 had reached 10% in mid-1973.

- Also, in the first 6 months of 1972, AN/SQS-26 ships in the Mediterranean Sea remained in port an average of 51% of the time. In the first 6 months of 1973, the average time in port had increased to 73% but ran as high as 86% during one of those months.[15]

- The final underlying factor of considerable long-term significance was the decline in interest regarding active sonar operations after the introduction of U.S. towed arrays. These arrays were not only successful in exercises but also in real-world encounters with Soviet submarines, the great majority of which were still noisy enough to present an attractive passive target. For exercises, special noisemakers installed in U.S. submarines provided Soviet-like acoustic targets to the towed array.

MAY 1973: COORDINATED OPERATIONS IN THE PACIFIC OCEAN

In May 1973, USS *Stein* (DE-1065), equipped with the SQS-26 (CX), participated in a coordinated operation with USS *Agerholm* (DD-826), equipped with the SQS-23 and a LAMPS helicopter, while both were in transit from Midway to Guam. An open-ocean encounter was planned with USS *Sailfish* (SS-572).

During a scouting mission along the *Sailfish's* known approach route, LAMPS made an initial electronic warfare support measure (ESM)

Chapter 9 — Fleet Performance

contact on the submarine's surface-search radar. After LAMPS closed range along the ESM bearing and made visual contact with the submarine's snorkel mast, it lowered altitude below 3000 feet and maintained a 10-kiloyard trail range so as to remain undetected.

Following this event, *Stein* confirmed a strong second convergence zone passive detection on the submarine at an estimated 60 nautical miles with a second annulus width estimated to be about 3 kiloyards. A merchant ship was also in the area, and passive contact via the first convergence zone was gained on both the submarine and the surface ship.

With the assistance of LAMPS, *Stein* acquired an active convergence zone contact on *Sailfish* at a range of approximately 33 nautical miles (convergence zone environmental conditions were considered to be marginal at this time). This contact was held for the next 4 hours with little difficulty. The mean convergence zone range during this period was 65.5 kiloyards, and the zone width was never more than 2.0 kiloyards. Convergence zone contact was held regardless of submarine operating depth.

During the same period, LAMPS was vectored out to *Stein's* datum to obtain MAD verification. *Agerholm* was vectored to the contact area, where it initially remained passive, but became active as the range to *Sailfish* closed.

Throughout this time, *Stein* used its passive sonar capability, along with its active convergence zone mode, to simultaneously track *Sailfish* and *Agerholm*. Several times, *Stein* vectored *Agerholm* back within sonar range of the target after *Agerholm* was unable to maintain contact.

The combination of the SQS-26 ship, LAMPS helicopter, and SQS-23 ship — all maintaining contact (each with its own sensor) for the entire exercise period — led the commander of Destroyer Squadron (DESRON) Five to comment as follows:

> The LAMPS/CZ team provided [a] long-range, surprise capability and complementary verification to allow early detection, continuous tracking, and prosecution of the submarine contact. It proved to be a most effective team.[16]

Chapter 9 — Fleet Performance

AUGUST 1973: SHAREM XVI (MD) CONVERGENCE ZONE RESULTS IN THE MEDITERRANEAN SEA

The Ship ASW Readiness/Effectiveness Measurement (SHAREM) XVI (MD) exercise was conducted from 27 August through 4 September 1973 in the Ionian Sea. The depth for convergence zone operation was marginal, which undoubtedly accounted for some of the missed opportunities.* Participants included *Belknap*, the on-site tactical commander (OTC); USS *Elmer Montgomery* (DE-1082); USS *Vreeland* (DE-1068); and USS *Lapon* (SSN 661). All surface units were equipped with the SQS-26, with *Elmer Montgomery* and *Vreeland* having CX models with steel domes. *Belknap* had an AXR model and had been fitted with a rubber dome window during the past year.

The exercise was considered "semifree" play in that the submarine had significant knowledge of the surface ship range, search plan, and intended movement. On the other hand, the surface ship expected a penetration attempt within a time window of approximately 1 hour, and it knew the general threat sector, although not the exact target bearing. The submarine was free to choose best evasion speed, aspect, and depth. Submarine depth actually varied from 50 to 500 feet, submarine speed from 5 to 20 knots, and aspect from bow to beam. Destroyer speed averaged 10 knots.

As expected, the rubber window ship, *Belknap*, had the most success. *Belknap* gained convergence zone contact in 41% (14/34) of the convergence zone detection opportunities presented by the target. The two steel dome ships gained convergence zone contact on only 15% (6/40) of the target opportunities. As stated above, some of the missed opportunities were probably due to a water depth that was insufficient for the full development of a convergence zone, and some of the zone widths were probably narrowed for the same reason. Detection ranges varied from 44.5 to 49.6 kiloyards. The mean observed annulus width was 3.7 kiloyards.[17]

---

*Computed depth excess from the convergence zone slide rule (the excess in depth over that which would result in the depth being on the borderline of blocking all rays from reaching the convergence zone) was evenly distributed from 50 to 220 fathoms. Two hundred fathoms is normally the minimum required depth excess for a fully developed convergence zone. Shipboard operators all too often forget about this restriction.

Chapter 9 — Fleet Performance

DECEMBER 1973: ANALYSIS OF ATLANTIC FLEET'S INTEGRATED ESCORT TACTICAL DEVELOPMENT PROGRAM

From 1971 through 1973, the Antisubmarine Warfare Force, Atlantic Fleet, ran the Integrated Escort Tactical Development Program, known by its short title as the Integrated Escort Program (IEP). The objectives of this effort were (1) to investigate the problems and capabilities of an integrated task force transiting open-ocean waters against an opposing submarine force and (2) to develop optimum tactics for encounters against such a submarine force.

In support of this effort, NUSC's Dr. David Williams (a member of my department) spent significant time in (1) planning exercises, (2) riding the ships as an observer, (3) assisting in the reconstruction of results, and (4) analyzing sonar performance. The information that Williams acquired would provide the first real understanding of ASW capabilities and limitations for the active sonars of that era in large-scale ASW exercises in the western North Atlantic. In addition, Williams recommended improvements in escort and task group tactics and in the modeling of escort/submarine encounters. (In 1973, I wrote a recently declassified technical memorandum on some of the implications of his analysis.[18])

The Atlantic Fleet operating areas off the East Coast of the United States were not suitable at that time for use of the long-range convergence zone and bottom bounce modes. The convergence zones were formed at ranges of 35 miles, incurring propagation losses too large to be tolerated by SQS-26 equipments with steel dome self-noise characteristics. Thirty miles was considered to be the upper limit for reliable convergence zone search with steel dome systems, and the rubber window had not yet been introduced to the Fleet. Moreover, bottom loss charts covered only portions of the North Atlantic. Even when bottom loss charts existed, the complexities of choosing the right depression angle and search window for bottom bounce operations required training beyond that received by most sonar operators.

The North Atlantic environment thus presented NUSC with the first opportunity to examine SQS-26 gains over predecessor sonar equipments when the *surface duct detection path* was used.

Chapter 9 — Fleet Performance

The performance of the SQS-26 in the surface duct mode had been a controversial topic. With early doubts about the effectiveness of bottom bounce and convergence zone modes, it was believed that the surface duct mode might have to be depended upon most of the time, at least in some areas. Critics asserted that (despite its larger aperture, lower frequency, and higher power) the SQS-26 as a surface duct sonar (1) would provide little or no gain over the performance of the older high-frequency systems, (2) would cost more, and (3) would broadcast the presence of the ship to a greater range. It was also thought that the "beaconing" effect would permit a submarine to avoid detection, as well as to target a surface escort and even perhaps the high-value carrier unit that the escort was protecting with a long-range cruise missile. (When the new FFG-7 escort class was in the planning stages in the early 1970's, this thinking contributed to the decision to mount a small, low-power, high-frequency sonar system (the SQS-56).) Other considerations for sonar selection concerned (1) the new emerging passive towed array systems and (2) the emphasis on minimizing ship cost.

The IEP included a series of task force exercises in which a "Blue" force — consisting of surface, subsurface, and air units — attempted to transit a deep-ocean area while opposed by an "Orange" force — consisting of submarines generally aided by air reconnaissance and other intelligence support. The duration of each exercise was about 1 week.

In the first four exercises, conducted through early 1973, the total number of sonars involved were as follows: 11 SQS-26's and 13 SQS-23's and 5 higher frequency sonars. These high-frequency systems included the SQS-41, the Canadian 503/504, and United Kingdom types 170 and 177. To enlarge the data sample, results from the contemporary Squeezeplay XI were included in some parts of the analysis.

NUSC's Williams established submarine detection-to-opportunity ratios for these exercises. To determine sonar performance independent of tactics, a "detection opportunity" was said to exist for the following conditions:

- The submarine was not at a bearing within the sector subtended by the escort's baffles.
- The escort sonar was operating in the proper mode and was covering a sector and range window that included the submarine.

Chapter 9 — Fleet Performance

- The sonar operator was not alerted to the submarine's location.

- The submarine was in one of four range bins relative to the predicted sonar range (PSR). Bin R1/2 included submarine ranges from 0 to PSR/2, bin R1 from PSR/2 to PSR, bin R2 from PSR to 2PSR, and bin R3 from 2PSR to 3PSR. Of course, detection opportunities were not equal in each bin, but gradually weakened with the increase in bin number. However, based on the experience of uncertainties involved in the prediction process, it was known that there would be a finite chance of detection, even in bin R3. The PSR was determined from the latest range prediction publications.

A feeling for how "unalerted" the operators were in these exercises can be conveyed by the fact that the average opportunity time amounted to 20 minutes per escort per week of steaming (a condition conforming to one description of naval warfare as consisting of long periods of boredom separated by short periods of extreme anxiety).

Figure 18 shows the sonar detection results for four exercises, each of which involved a 10-knot task force transit of about 1-week duration, with one or more high-value units being protected by destroyers, submarines, and VP aircraft. The formations were typically quite spread out, about 40 to 150 miles across. Williams, who participated both in the planning and analysis of these exercises, was accompanied by other NUSC observers aboard selected units. Consequently, NUSC obtained a fairly accurate idea of what occurred with respect to sonar performance.

In figure 18, the SQS-26 performance was separated from that of the older lower power and higher frequency) sonars. Both types of systems were represented in the exercise — 11 ships were equipped with the SQS-26 and 13 ships with the older sonars. The detection opportunities were divided into bins corresponding to the closest point of approach (CPA) for each opportunity. The SQS-26 clearly obtained substantially longer detection ranges. No detections on the older systems were obtained beyond the 10-kiloyard lateral range bin, whereas SQS-26 detections were obtained out to the 40-kiloyard bin, with a significant detection-to-opportunity ratio. Furthermore, the SQS-26 missed fewer opportunities than did the older systems. However, because 7 of the 15 SQS-26 detections and 2 of the 4 detections for the older systems were

Chapter 9 — Fleet Performance

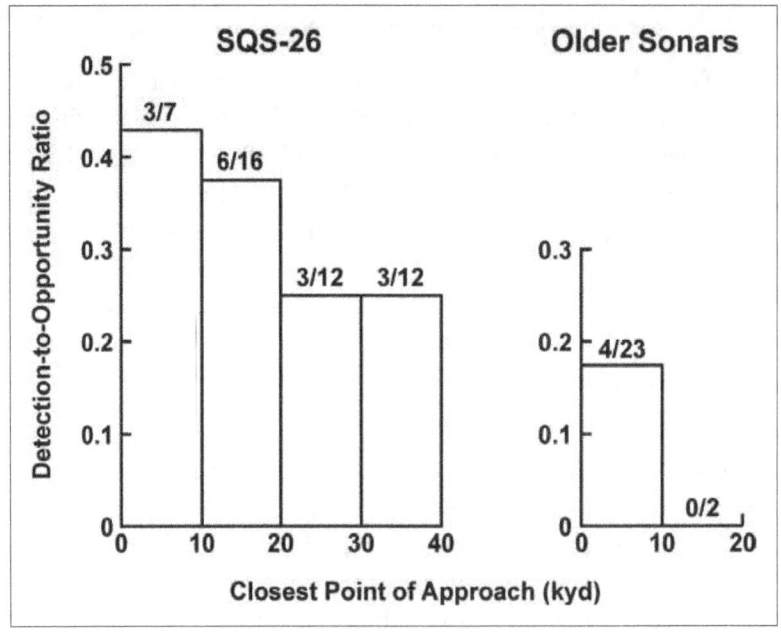

Note: In this sample, the older systems obtained no detections beyond 10 kiloyards. The SQS-26, however, obtained detections between 30 and 40 kiloyards, with a reasonable detection-to-opportunity ratio for these ranges.

Figure 18. Detection-to-Opportunity Ratio Versus Closest Point of Approach for Both SQS-26 and Older Sonars

converted into attacks, there was no significant difference in the probability of converting a detection into an attack once a detection was made. That is, for both types of sonars, about half of the detections were converted into attacks.

Looking at results from all the IEP escort exercises, in addition to some added data from the Squeezeplay exercises, the SQS-26 obtained substantially longer detection ranges (averaging 19 kiloyards) than the detection ranges achieved by the older high-frequency systems (averaging 6 kiloyards). This result was not too far from the predicted average detection ranges for the two groups of systems in these environments — 22 kiloyards for the SQS-26 and 6 kiloyards for the older systems. While the expected improvement in detection range of somewhat more than a factor of three over the older sonars seemed to be

Chapter 9 — Fleet Performance

borne out by the exercise results, the average range statistic by itself is of limited significance because the missed opportunities are not included.

Another way to present the results is by normalizing the detection and opportunity ranges to the expected 50% detection range, as shown in figure 19. This normalization addresses both environmental and basic range capability differences to the extent that performance prediction techniques allow. The plot, of course, does not reflect the average 3:1 differences in the predicted detection ranges. To increase the sample size, the results from Squeezeplay XI, which involved a 1-day ASW search exercise, are included. In that exercise, the opportunities per hour were relatively high.

For some reason, the SQS-26 system seems to perform considerably better than the older sonars in capitalizing on opportunities, even after normalizing out its basic factor of three in detection range advantage. It should be noted that in the 0.5 to 1.0 bin, where one would expect at least half of the opportunities to be converted to detections (since, by definition, the predicted detection range corresponds to a 50% probability of detection), the SQS-26 made only 7 detections from the 20 opportunities in the bin (for a 0.35 probability). Yet, in the 0 to 0.5 bin, six out of seven contacts were made for a 0.86 probability, which was astonishing, even after consideration of the small sample size.

The older high-frequency systems, on the other hand, detected only 2 of 16 opportunities in the 0.5 to 1 bin (for a 0.13 probability) and 3 of 9 in the 0 to 0.5 bin (for a 0.33 probability). Even allowing for sampling errors, the far lower than expected probability for the high-frequency systems is significant. Because a greater priority was placed on the maintenance, manning, and training programs for the SQS-26 than for the older systems, it is possible that this result is due to the better utilization of the SQS-26 equipment.

While the North Atlantic results presented herein indicate that the SQS-26 did well on the opportunities received, one often-made assertion was that the pinging itself results in a degradation in detection effectiveness by producing a *decrease in opportunities* that would not be reflected in the plot shown here. It is alleged that the submarine was able to stay

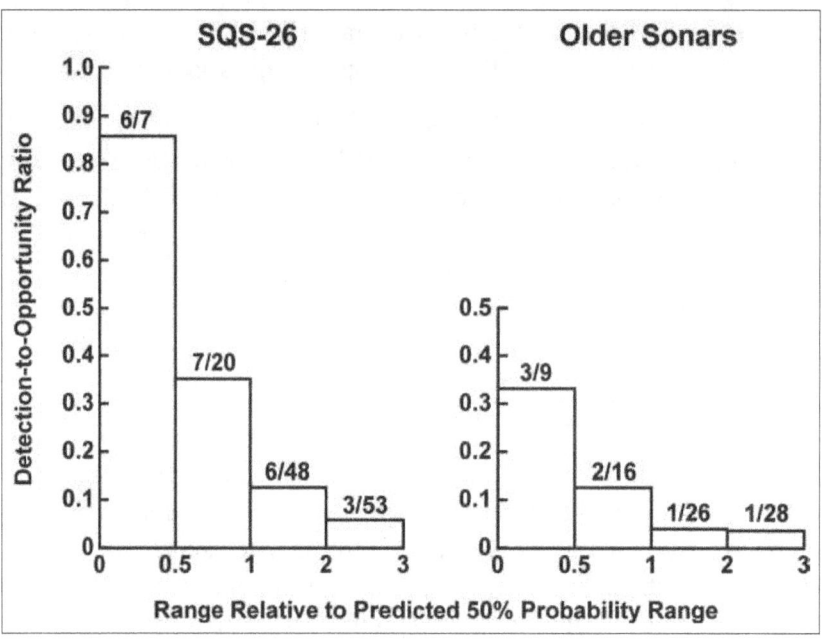

Note: This plot does not reflect the 3:1 advantage in the 50% probability range for the SQS-26.

Figure 19. Detection-to-Opportunity Ratio Versus Range Normalized to the 50% Probability Detection Range

away from the escort's pinging sonar, thereby markedly reducing detection opportunities and thus detections.

An initial effort to investigate this hypothesis was made by examination of the submarine CPA distribution during periods when the active sonar was turned off. No significant differences could be found between the CPA distributions for active and passive sonar operation, suggesting that the submarine does not successfully utilize the escort's sonar pinging for avoidance. Another way to address this issue is to investigate whether there are fewer close-in opportunities for active detection than would be expected on a random submarine density basis.

In table 3, the hours of detection opportunities on submarines within a given normalized range are tabulated versus what would be expected on a random density basis. If the submarines are randomly distributed, then the opportunity hours should be proportional in a range sample to the area enclosed by the corresponding normalized range; that is, a

Chapter 9 — Fleet Performance

normalized range of three contains on the average nine times the area enclosed by a normalized range of one. It is seen that the distribution of opportunity time is (within sampling error) identical to what would be expected on a random submarine density basis; that is, the opportunity time is very nearly proportional to the area enclosed by the corresponding normalized range, indicating that the submarine density per square mile is the same within the expected detection range as it is well beyond that range. Thus, there is no evidence that the submarines spent any more time at the longer ranges than they would have if the escort ship were completely silent.

Table 3. Distribution of Opportunity Hours as a Function of Range Normalized to 50% Probability Detection Range*

Range Normalized to 50% Predicted Range	Opportunity Time Within Normalized Range		
	Hours	% of Total	% Expected for Random Density
0.5	1.6	4	3
1.0	5.4	15	11
2.0	17.5	49	44
3.0	35.3	100	100

*This table suggests that the submarines are not successful in avoiding the active sonar of the escorts. Otherwise, there would be a markedly greater percent of opportunity hours at the longer normalized ranges than would occur for a random submarine density.

DECEMBER 1974: OBSERVATIONS OF INCORRECT SYSTEM OPERATION

Feedback from Fleet exercises up through December 1974 was obtained on equipment operation, as well as on detection results. Errors in the operation of the equipment, observed on more than one occasion, are described next.

Power Settings

Power was sometimes reduced in the mistaken notion that this condition would also reduce reverberation and thereby improve performance. What actually happened was that the echo level was reduced as much as reverberation, with the result that no change was made in the signal-to-reverberation ratio. On the other hand, if a significant component of noise existed in the background, the signal-to-background ratio would be decreased with power reduction, thereby weakening detection performance.

Depression Angle for Convergence Zone Search

Sometimes the wrong depression angle was used to set up the convergence zone equipment mode. During the period in question, a depression angle of 5° was recommended in the operating doctrine, but the actual angles used varied from 0° to 8°.

Random Mode Settings

During transit situations, random modes were often used, apparently on the theory that a lack of detections on the existing mode suggested that it was the wrong mode and that another selection would offer an improvement. There was, of course, no way in search operations to know whether or not the negative results were just simply due to no available target opportunities. The recommended mode from the doctrine would be expected to provide the best chance of detection if there was a target opportunity.

Passive and Active Operation

Some commands thought it best to alternate passive and active operations on the theory that this approach would make it more difficult for the submarine to locate the surface ship. The problem with this strategy was that the active system needed to operate for only a small fraction of time before the submarine could discern the surface ship bearing. When a submarine target was so quiet that passive detection was not possible, then it was better for the surface ship to be active 100% of the time. However, if the target was expected to be noisy enough that a passive detection could be acquired, there was usually no point in employing the active system, at least prior to detection.

Sound Velocity Settings

Sound velocity for a convergence zone mode was set as the surface velocity rather than as average velocity (appropriate for travel through almost the whole depth of the water). This choice could result in errors of several thousand yards of range that would cause a serious problem when the attempt was made to vector out another ship or aircraft to the convergence zone datum.

OCTOBER 1975: LAMPS III TESTING IN THE NORTH ATLANTIC

In 1975, the last of the special experiments with the SQS-26 took place in the North Atlantic, where convergence zone detections could be expected to occur beyond 70 kiloyards. Testing was conducted with *Connole* (now designated FF-1056) in the role of support to LAMPS III.

Connole, equipped with an SQS-26 (CX) and a rubber dome window, carried a LAMPS experimental helo that was to simulate a LAMPS III. During the tests, it was proven that *Connole* could reliably detect a below-layer submarine beyond 70 kiloyards, as well as hold contact while LAMPS flew to the location and delivered a simulated attack. Holding contact could be accomplished either by maneuvering to parallel the contact in the convergence zone or by closing and reacquiring contact in the bottom bounce mode. Acquisition was made on each of the runs, with a target motion analysis (TMA) solution calculated as the contact transited the zone, illustrating that bearing and range accuracy was sufficient to determine submarine course and speed.

APRIL 1976: *CONNOLE* AND THE ASW SQUADRON IN THE MEDITERRANEAN SEA

In April 1976, the ASW Squadron was deployed to the Mediterranean Sea. The idea of forming an ASW squadron had developed as the result of early discussions between the Commander, Naval Surface Forces, Atlantic Fleet (COMNAVSURFLANT), and NUSC.

The commanding officer at NUSC, Captain Milton McFarland, had proposed that NUSC meet quarterly with the COMNAVSURFLANT staff to obtain Fleet input on problems and to present information on emerging programs at NUSC. At one of those meetings (which also included NUSC technical director Harold Nash), I gave a formal presentation on the concept of equipping *Connole* with the latest in ASW equipment to demonstrate what could be done against the Soviet submarine threat in an important area like the Mediterranean. Such equipment would include the most recent (available) improvements in both passive and active sonar. Commander Richard F. Pittenger, the staff ASW officer, and I were in frequent contact to follow up on details of what would be done. The *Connole* proposal grew into the concept of forming a Mediterranean ASW Squadron composed of SQS-26 ships.

Figure 20 shows the ships of the ASW Squadron at their berth in Naples, Italy: USS *Moinester* (FF-1097) and *Connole* (FF-1056) of the *Knox* class, *Koelsch* (FF-1049) and *Voge* (FF-1047) of the *Garcia* class, and *McCloy* (FF-1038), which was one of the two ships of the *Bronstein* class. *Moinester* and *Connole* were both equipped with rubber dome windows, towed arrays, and LAMPS I helicopters. All ships were equipped with SQS-26 sonars.

Fortunately for NUSC, Commander Pittenger — with his unique understanding of modern sonar capabilities from past assignments — was moved from his COMNAVSURFLANT position to the command of the *Connole* in the summer of 1976. Earlier, in 1965, he had obtained a Master of Science degree in physics (underwater acoustics) from the Naval Post Graduate School in Monterey, California. Next, at the Naval War College in Newport, Rhode Island, he wrote a thesis on the history of sonar, after which he was assigned to a tour at COMNAVSURFLANT, where NUSC had frequent interaction with him. He was later promoted to rear admiral as Oceanographer of the Navy.

One of many highlights during Pittenger's deployment with the ASW Squadron was the unalerted active detection in the convergence zone of a Soviet *Echo-2* class submarine. This event occurred while *Connole* was screening an aircraft carrier, whose chief of staff rode out to the datum in a helicopter to see what was going on with this contact that *Connole* was supposedly holding at an over-the-horizon range. Just as he arrived at the datum, the submarine raised its periscope, providing convincing evidence that the *Connole* had acquired contact on a real submarine. As a result, the *Connole*, which was about to be directed to break off sonar operations and occupy a carrier lifeguard station, was allowed to continue with its tracking of the submarine. Commander Pittenger's reaction was as follows: "He [the OTC] ordered us to stay in contact, which, of course, we did with glee." *Connole's* total active contact holding time was 13 hours.[19]

Moinester, *Connole*, *Koelsch*, *Voge*, and *McCloy* are shown from left to right.

Figure 20. ASW Squadron at Naples in 1976

FINAL REPORT ON FLEET RESULTS WITH THE SQS-26

The January 1976 report on Fleet results documented the final efforts on the NUSC development of an operating doctrine program that had been based on close feedback from the Fleet regarding their experiences in operating the SQS-26 equipment in all propagation path modes, but most particularly in the convergence zone mode.[20]

By the end of 1975, the Fleet showed little interest in the active mode of the SQS-26 due to the effectiveness of passive towed arrays. With the reduction in operating days at sea, it was only natural for the Fleet to concentrate on what seemed, in general, to offer the most success on the large population of noisy Soviet nuclear submarines.

The period between 1974 and 1989 was the era of the passive system in ASW, as discussed in chapter 8. Before and after that time, active sonar was required, especially when it became necessary to cope with the silent submarines that appeared in significant numbers around 1990.

Chapter 9 — Fleet Performance

SIGNIFICANCE OF FLEET PERFORMANCE OBSERVATIONS

The Fleet reporting and analysis program managed by NUSC yielded a total of 471 convergence zone contacts between June 1970 and December 1974. Most were obtained against surface ships because these were the most numerous targets, but about 25% were obtained against submarines. The reflectivity of a surface ship is close enough to that of a shallow submarine to say that the surface ship detection events provided worthwhile practice for detecting and holding contact on a periscope-depth submarine.

Actively detecting and holding contact on real submarines in the exercises was also demonstrated sufficiently to say that the SQS-26 with a rubber dome window had a strong capability for submarine detection in the convergence zones of the world's oceans (when environmental conditions permitted such zones to exist). There was also repeated evidence that the detecting ship could vector out an aircraft or another ship to successfully reacquire and attack the target.

The concern from the early days of the SQS-26 program that the sonar system might prove to be no more capable than its SQS-23 predecessor (but also more costly and less reliable) turned out to be unfounded. The new bottom bounce and convergence zone long-range modes — especially the convergence zone mode — demonstrated a unique capability against both deep and shallow targets in adverse near-surface thermal conditions. Where conditions for bottom bounce and convergence zone modes did not exist, the surface duct mode of the SQS-26 sonar (when available) provided clear superiority over the same mode in the older sonars.

The most dramatic difference shown by the SQS-26 was in the Mediterranean Sea. When Captain Kelly from the Sixth Fleet contacted NUSL in 1968, he explained that his ASW capabilities in the face of a rapidly growing Soviet buildup were essentially nonexistent. In 3 years, the Soviets had increased their submarine presence in the Mediterranean Sea by nearly a factor of 10, but U.S. submarine detection ranges with pre-SQS-26 sonar were little more than 1000 yards in the severe thermal gradients characteristic of the warm months of the year — even against periscope-depth submarines. With the favorable convergence zone and bottom bounce conditions on deep sound paths, along with the capabilities of the SQS-26 to exploit those paths, the Fleet's submarine detection situation in the Mediterranean was completely turned around.

CHAPTER 10
CONCLUSIONS

ACCOMPLISHMENTS OF THE SQS-26 PROGRAM

The years of experimentation with the SQS-26, along with the effort undertaken in associated support programs, yielded a rich harvest of information relevant to long-range active sonar — information that was largely unknown when the NUSL program was originally initiated in 1955.

As a result of the SQS-26 program, knowledge was greatly enriched in the following areas:

- The attenuation of sound in sea water;
- Surface, biological, and bottom backscattering;
- The reflectivity of the ocean bottom;
- FM pulse signal processing behavior under conditions of "energy splitting loss" in the medium; and
- Doppler processing against real echoes.

Design technology was also enhanced for the following sonar equipment:

- High-power transducer arrays;
- High-power sonar transmitters feeding the array elements;
- Transmitting and receiving beams electrically steered in both azimuth and elevation through a cylindrical array;
- Displays receiving echoes over all sonar paths; and
- Sonar dome structures, with the rubber dome window development accounting for the single most noteworthy gain.

Much was also learned on SQS-26-type sonar about how to predict shipboard performance, train the Fleet in the tactical use of equipment,

Chapter 10 — Conclusions

obtain feedback from Fleet operations, and develop the information needed to operate the equipment in any given environment.

Finally, valuable information and skills were acquired on the development of methods for quantitatively testing and analyzing the performance of a multiple-path, long-range sonar.

Although the characterization of SQS-26 development in the summary document prepared by NUSL/NUSC's* William Downes in 1971 may appear boastful to anyone not familiar with the effort, his conclusion would easily be supported by an objective review of the program's accomplishments:

> [NUSC] . . . successfully conducted what had probably been the most extensive and fruitful program of *sea tests* in the history of U.S. sonar system testing.[1]

Downes' focus on sea tests in this statement was appropriate since NUSC activities during the development of the SQS-26 system were almost entirely motivated by information derived from tests at sea.

The foregoing chapters have largely provided a memoir of my own involvement in the SQS-26 development program. For this reason, much of what would be of interest regarding *hardware development* (in which EDO, GE, and NUSC were all heavily invested, but in which I had only minimum involvement) has been excluded. I do not want to appear to slight the many valuable contributions of NUSC and the two contractors in meeting the formidable challenge of solving the many problems encountered in developing the hardware. However, the budgeted scope of this documentation effort made it impractical to adequately treat those areas.

CONTRIBUTIONS TO PROGRAM SUCCESS

You might ask, Was the overall project a success? In the process of continually defending the program during its 20 years of development, I found that critics often felt that the success or failure of a project was determined by comparing actual capabilities achieved with early

---

*Although use of the "NUSL/NUSC" designation would be appropriate in many instances in this chapter, it has been shortened to "NUSC" in the interest of brevity.

Chapter 10 — Conclusions

promises. If a project did not fully measure up to the early performance estimates, it was argued that the project had failed. Yet, because those who made early projections could not possibly foresee the many problems that would be encountered along the way, I felt that this definition of success was highly inadequate. What should have been asked was the following question: Did the new system provide a sufficient improvement over its predecessor system to justify the commitment and expense? A review of the performance just recounted in chapter 9 provides a strong argument that the unique "over-the-horizon" detection and attack follow-up capabilities demonstrated by the SQS-26 relative to its SQS-23 predecessor indeed made the SQS-26 a worthwhile investment that well justified the additional time and cost.

Now you might wonder, What *specifically* contributed to the success of such a complex, lengthy program as the SQS-26? Although the leadership provided by William Downes was crucial, as was the experience and talent of the team he assembled, other important influences should not be overlooked:

- *The international situation*: Continual displays of enhanced military capabilities by foreign powers and their frequent confrontations with the United States provided a convincing rationale for maintaining a strong and technically advanced U.S. Navy in the ASW area. This situation was a major contributor to the continuing support of the SQS-26 development program.

- *The Navy laboratory system*: The Navy laboratories played a key role, both at NUSC and elsewhere,* by providing essential support for a unique combination of sonar-related activities that included the following: applied research, systems engineering, hardware design, testing, specialized facilities, life-cycle involvement, quick-reaction capabilities, intelligence awareness, and Fleet access. Elements of this support were also obtained from liaison with government laboratories in Canada and the United Kingdom, as well as in La Spezia (Italy), the site of NATO's SACLANT Undersea Research Center. Support of a different type was received from the Naval Oceanographic

---

*See chapter 2, Post-War Research on Long-Range Sound Paths.

Chapter 10 — Conclusions

Office, which conducted survey activities and provided oceanographic data files that were of key importance.*

- *Upper echelon management support*: Excellent direction and funding support from CNO and NAVSEA (formerly BuShips) were provided with little interference in NUSC technical decisions.†
- *Policy on contracting with private industry*: The policy on contracting with industry for hardware development and production, with technical guidance provided by NUSC, worked out extremely well.‡

The following pages further discuss how all these influences affected the SQS-26 program.

INTERNATIONAL SITUATION

A continual series of international incidents, which resulted in high military tensions for some 5 decades after the end of World War II in 1945, were conducive to the maintenance of a strong defense policy. This effect, in turn, led to the continuing financial and requirements support of the SQS-26 project during its lifetime. Several of the events that occurred during what was termed the "Cold War" are described next:

- In 1948, the Soviets began a 2-year blockade of Berlin.
- In 1950, the North Koreans attacked South Korea. U.S. involvement in the Korean conflict, which was later to bring in the Chinese on the side of the North Koreans, would last for 3 years.
- In 1956, the Soviets invaded Hungary to maintain Communist control of that country.
- In 1962, the Soviets attempted to install missiles in Cuba, provoking the "Cuban missile crisis" that ended with President Kennedy imposing a naval quarantine of Cuba.

---
*See chapter 6, The Marine Geophysical Survey Program.
†See, for example, chapter 3, Melding the NUSL Sonar and CNO Scout Ship Concepts.
‡See chapter 3, Procurement Plans.

- In 1965, the United States began to build up forces in South Vietnam to oppose a Communist takeover by the Viet Cong in North Vietnam. This situation ended in 1975, with U.S. withdrawal and the North Vietnamese capture of Saigon.

During all this time (from post-World War II to 1991), the Soviets were constructing 727 submarines, at an average building rate of 16 per year. This situation ensured continuing U.S. interest in improving Navy ASW capabilities, including support for the SQS-26 program.

NAVY LABORATORY SYSTEM

The Navy laboratory system has a unique combination of capabilities that taken together are not shared by any commercial organizations or university laboratories. These capabilities are described next.

Applied Research

The research activities of the Navy laboratories largely respond to a "requirements pull" rather than to a "research push." Although NRL probably provides more emphasis on the research push than do the other laboratories, it has also demonstrated an ability to respond to the requirements pull.

The SQS-26 system concept was based on a "directed research" push in propagation measurements and long-range, echo-ranging techniques at NUSL, NRL, and NEL. During development of the SQS-26 system, supporting research information would be contributed in selected areas by government laboratories in Canada and the United Kingdom and by the NATO laboratory in La Spezia.

From both inside and outside the United States, government laboratory organizations provided specific research support (as related in foregoing chapters) in the following areas: transducer design, shallow-water echo-ranging, signal processing, attenuation, convergence zone propagation loss, biological reverberation, bottom reverberation, sea surface backscattering, surface loss, and bottom loss. Although the nature of this support was often due to information obtained from informal liaison, the Naval Oceanographic Office provided specialized survey support, mainly on bottom loss. However, their existing data files on bathythermographs, deep sound speed profiles, salinity, bottom

Chapter 10 — Conclusions

topography, sea state, and surface temperatures were also immensely important to the SQS-26 effort.

Systems Engineering

Along with developing its own unique capabilities, NUSL also kept up with practices that were occurring in the commercial sector, such as the new emphasis on systems engineering (see chapter 3).* The systems engineering functions, which were fully supported by the Navy laboratory system as relevant to SQS-26 development, involved the following processes:

- Applying the results of directed research to the conceptual design and development of the SQS-26 and maintaining a close association with the other research organizations throughout the process.

- Maintaining close contact with those involved in SQS-26 equipment development and continuously monitoring SQS-26 technical difficulties.

- Amending objectives and plans as required.

- Organizing field trials as needed.

- Planning tests and evaluating their results;

- Following the in-service performance of first installations and reacting to the problems associated with these new systems as they entered Fleet service.

- Participating in the evaluation of the system and determining its military worth.

Hardware Design

The Navy laboratories must always maintain a technical design capability in militarily important fields that are of limited interest to the commercial sector. One example of this is found in the area of low-frequency transducer design for submarine detection applications.

---

*I eventually headed the systems engineering organization (then called "systems analysis"), which became a department under the technical director. Before that, I had assumed the same responsibilities under William Downes in Surface Ship Sonar. (See page 52 for a discussion of systems engineering.)

Chapter 10 — Conclusions

The emphasis on strengthening its own design capabilities ensures that the Navy not only has the background required to become a "smart buyer" of commercial products, but also has the skills necessary to supply information that will help industry to work out design problems — a particularly important advantage for the naval community when a commercial company is developing military products.

Testing Expertise

Navy laboratories have traditionally been relied upon to test products for which the Navy has contracted. New London's NUSC, in particular, excelled in shipboard sonar sea test capabilities as the result of extensive experience that was acquired as far back as World War II. Over the years, this laboratory had built up a reservoir of skilled and talented engineers and technicians whose depth of expertise allowed them to meet the demands of the SQS-26 development program, which depended heavily on the reliability of the sea test effort.

Specialized Facilities

NUSC maintained sonar test instrumentation, well-equipped test barges, test tanks, measurement basins, and shop facilities for the manufacture of specialized test hardware. For example, the measurement barge at Dodge Pond in nearby Niantic, Connecticut, was available to make initial tests on the SQS-26 array and transmitter drivers.

The Navy laboratories, in general, also had extensive sonar documents libraries that contained classified and unclassified publications going back to the early part of the 20th century. The library facilities provided key support for obtaining the technical information that was continually required during the life cycle of the SQS-26 project.

Life Cycle Involvement

The involvement of the Navy laboratories over the lifetime of a sonar system results in services that are unique and in personnel that have the specialized perspective and know-how for handling the practical problems that arise during the design of a new product. In particular, NUSC provided testing expertise for the new SQS-26 production installations, troubleshooting for problems that the ship could not solve, and liaison with the contractor regarding required corrective action. All

Chapter 10 — Conclusions

such activities furnished valuable feedback that resulted in improvements to the original design.

Quick-Reaction Capability

NUSC was able to react quickly when it became evident to the Navy that more effort was required in selected program areas than was originally expected. For instance, services were immediately offered when it was necessary to acquire more information on the bottom environment, to assist in training personnel, to prepare operating manuals on performance prediction and operating doctrine, to troubleshoot difficulties with the hardware design, and so on.

The organization of the effort that would address each of these problems was typically accomplished in a matter of weeks during the development phase of the program, without the added burden of time-consuming paperwork for requests for proposals, contractor responses, management plans, etc. Although contracting was required in certain instances, it would be arranged to proceed in parallel with the in-house effort.

Intelligence Awareness

The Navy laboratories have a special office that is responsible for maintaining liaison with the intelligence community. This arrangement ensured that NUSC was kept abreast of enemy capabilities during the development of the original conceptual design of the SQS-26, as well as during the improvement efforts undertaken for that design.

Access to the Fleet

The Navy laboratories have easy access to the Fleet ships and headquarters organizations.* Laboratory personnel are welcome to attend briefings and to ride ships on ASW exercises to observe sonar performance, make special checks on equipment readiness, receive feedback on Fleet problems, and provide advice on sonar search and localization tactics. These interactions are invaluable for the formulation

---

*See chapter 9 for an outstanding example of laboratory/Fleet cooperation under the section entitled May 1968: Feasibility Studies of Convergence Zone Applications in the Mediterranean Sea.

of training programs, equipment modifications, and the design of new sonar equipment.

The laboratories also have convenient access to U.S. Navy officers assigned to the laboratory staff for a tour of duty. These officers typically have not only had ASW Fleet assignments, but have often acquired graduate education in acoustics and attended the Naval Postgraduate School. Such open access was important both before and during the SQS-26 development program.

UPPER ECHELON MANAGEMENT SUPPORT

Although the management and distribution of funding for a laboratory's projects are handled directly by NAVSEA, the naval laboratories prefer to think of themselves as having an independent "conscience," which would permit them to reject or modify assignments that do not appear to be in the Navy's best interests. However, when a laboratory exerts such independence, NAVSEA could conceivably choose to work with commercial contractors who are more responsive to its wishes.

On the SQS-26 program, relations between NUSC and NAVSEA were very good. The main point of contact for system hardware development and testing matters at NAVSEA was Elmer Landers and for operating doctrine and training was Paul Tiedeman. No technical recommendations that made sense economically and fit into the development schedule were ever turned down, and there was no interference when it came to technical support. This relationship was crucial to the success of the SQS-26 development program.

POLICY ON CONTRACTING WITH PRIVATE INDUSTRY

There was always understandable pressure to contract out to private industry what work it could more efficiently undertake. While contractors lacked many of the advantages of the Navy laboratories, there were other areas where they possessed unique capabilities.

For hardware development leading to the production of a system as large and complex as the SQS-26, the contractors were far better equipped than was NUSC, which had no desire to compete with them for this type of work. Throughout the project's life, good communication

Chapter 10 — Conclusions

was maintained with both EDO and GE. In the end, EDO manufactured 18 SQS-26 (BX) systems and GE furnished 69 SQS-26, SQS-26 (AXR), and SQS-26 (CX) systems.

In the area of active signal processing (where NUSC had little expertise), it became advantageous for NAVSEA to contract with Tracor for specialized support. Tracor also furnished other services for which it was particularly skilled.

The contracting policy worked extremely well and helped to ensure the overall success of the SQS-26 project.

SUMMARY OF CONTRIBUTIONS TO PROGRAM SUCCESS

There were many influences contributing to the success of the SQS-26 system, which, of course, began with the leadership of NUSL's William Downes, who was responsible for assembling the program's capable development team. But there were also other factors to consider. The international tensions with the Soviets that persisted throughout the project's development, along with their aggressive submarine-building program and worldwide submarine deployment, ensured national interest in maintaining a strong ASW capability. The Navy laboratory system, both in this country and among foreign allies, provided unique capabilities not elsewhere available for applied research, systems engineering, hardware design, testing, specialized facilities, life cycle involvement, quick-reaction time, intelligence awareness, and access to the Fleet. Management from the upper echelons was competent and supportive. And finally, the philosophy of contracting out hardware production and specialized support projects worked extremely well.

TRENDS IN ASW BEYOND 1975

The SQS-26 development program ended in 1975. Influences that diminished Navy interest in surface ship active sonar in the early 1970's have already been alluded to in chapter 8. Dominant among these were the radiated noise characteristics of Soviet submarines and the towed array sonar techniques on U.S. Navy surface ships that permitted long-range passive detections and attack follow-up without the need for active sonar. Beyond 1975 through the end of the century, trends in the Soviet

Chapter 10 — Conclusions

submarine threat continued to influence attitudes toward the development and use of active sonar, which will be briefly reviewed next.

In the early 1960's, it was mistakenly assumed that the silencing successes achieved by U.S. submarines in the 1961 *Thresher/Permit* class would be duplicated by the Soviets, who had seemed capable of matching U.S. military systems with only a few years of delay. The first Soviet nuclear submarine, *November* SSN, was launched in 1958, some 4 years after the first U.S. nuclear submarine. By 1968, a total of 55 Soviet nuclear submarines had been deployed: 8 *Hotel* SSBNs, 34 *Echo* SSGNs, and 13 *November* SSNs (jointly referred to as HENs). The HENs turned out to be extremely noisy and easily detectable by U.S. submarine sonar arrays, the sound surveillance undersea system (SOSUS) fixed system network, and aircraft sonobuoys. More good news was to come with the beginning of the Soviet deployment in 1968 of the second-generation *Charlie/Victor/Yankee* class — the CVYs. These were quieter, but still noisy enough to be extremely vulnerable to passive sensors. By the early 1970's, the U.S. towed arrays being introduced experimentally in surface ships had a large population of Soviet HENs and CVYs for passive targeting. In 1978, production towed arrays started to become available to U.S. frigates with the introduction of the SQR-18 on USS *Joseph Hewes* (FF-1078). In 1982, even more capable towed arrays entered the U.S. Fleet when the SQR-19 was installed on the destroyer USS *Moosbrugger* (DD-980).

The 20-year period from 1960 to 1980 has been referred to by Dr. Owen R. Cote (Associate Director, MIT Security Studies Program) as the "happy time" in ASW, when the U.S. Navy relied primarily on passive acoustics to detect Soviet nuclear submarines, while the Soviets were unable to passively detect U.S. nuclear submarines.[2] However, Dr. Anatoly V. Kuteinikov — head of the Soviet Malachite Central Design Bureau in St. Petersburg and responsible for overseeing nuclear attack submarine construction — has stated that in the mid-1960's the Soviets were already beginning to address submarine silencing.[3]

Robert J. Murray, a former Defense Department official, has discussed the Soviet quieting trend. The first *Victor III* joined the Soviet Fleet in 1972, with low noise levels equal to those of the U.S. *Thresher/Permit* class. Murray credited the improved Soviet *Akula*, appearing in 1991, with being quieter than its U.S. contemporary, the *Los*

Chapter 10 — Conclusions

Angeles class. He attributed this quieting achievement to a combination of "the skill of Russian scientists and engineers, in part to Western technology illicitly acquired, and in part to help from two convicted American spies, John Walker and Jerry Whitworth, who for many years sold U.S. secrets to the Soviet Union."[4]

It took time for the quiet submarines to appear in the Soviet Fleet in significant numbers. J. Richard Hill wrote that passive sonar on surface ships was considered to be the primary method of submarine detection in the 15-year period from 1974 to 1989.[5] By 1986, however, signs that the reign of passive sonar was in decline were being recognized by U.S. experts. Norman Polmar, an expert on Soviet capabilities, stated the following in 1986:

> Interesting aspects [of recent Soviet developments] include quieting . . . their boats are getting quieter. They're getting quieter at a faster rate than our intelligence community predicted . . . . As boats get quieter, we may reach a point in the not-too-distant future where the only way we can detect the other guy, because he's so quiet, is [by] going active.[6]

In 1989, William D. O'Neil, former Assistant Deputy Under Secretary of Defense (Naval Warfare), acknowledged that with the reduction in radiated noise levels achieved by the new Soviet submarines, passive sensor ranges were reduced to only 5% of those experienced in the late 1970's.[7] Perhaps the first tangible reaction of the Navy to the decline in passive effectiveness was the elimination in 1992 of the SQR-19 towed array from the DDG-79 Flight IIA specification.[8] In 1994, Polmar credited Soviet source Sergey Pitchkin with the following quote: "The Americans lose *Oscar II* [a Soviet SSGN] immediately after the submarine puts out into the ocean!"[9] In 1997, the Naval Studies Board of the National Research Council wrote that "passive detection ranges for these low-speed [low-noise] modern submarines have shrunk from hundreds of kilometers to only a few kilometers."[10]

With the foregoing prospects affecting the future of passive sonar, one might think that there would be a resurgence of development activity in active sonar. This event, however, did not occur for a number of reasons:

Chapter 10 — Conclusions

- There has been a tendency in the scientific and at-sea operational communities to believe that the passive techniques that have been the mainstay of ASW for so long can somehow be revived if the appropriate signal processing, hydrophone arrays, and tactics are used.

- With little interest or practice in active sonar during the period from 1974 to 1989, the ranks of those knowledgeable in its use, both in the Fleet and in the development community, have thinned considerably.

- ASW no longer has a high priority on surface ships, with the understandable emphasis in the last decade on land-attack missiles and anti-air warfare (AAW) defense. With no U.S. Navy combat actions involving hostile submarines since 1945, it can be seen why ASW does not command the attention it once received.

- The Soviet threat, which had been responsible for keeping ASW funding at a high level since 1950, all but disappeared with the end of the Cold War in 1991. No comparable alternative threat is yet evident. Unless the ASW threat measurably changes, it is hard to imagine reviving the degree of emphasis on active sonar development and training that existed from 1945 through the early 1970's.

CHAPTER 11

ANNOTATED ENDNOTES*

CHAPTER 1: INTRODUCTION
Pages 1–4

1. J. Merrill and L. D. Wyld, *Meeting the Submarine Challenge: A Short History of the Naval Underwater Systems Center,* United States Government Printing Office, 1997, pp. 64-65 (UNCLASSIFIED).

2. Maurice Ewing's wartime research group at Woods Hole studied the convergence zone propagation path both theoretically and experimentally; see M. Ewing, "Interim Report No. 1," Woods Hole Oceanographic Institution, Woods Hole, MA, 25 August 1945 (UNCLASSIFIED); also see M. Ewing and J. L. Worzel, *Geological Society of America*, Memoir 27, October 1948 (UNCLASSIFIED).

CHAPTER 2: HISTORICAL BACKGROUND
Pages 5–18

1. F. V. Hunt, *Electroacoustics,* American Institute of Physics, New York, 1982, pp. 46-52.

2. E. Klein, "Notes on Underwater Sound Research and Applications Before 1939," ONR Research Report ACR-135, Office of Naval Research, September 1967, p. 7 (UNCLASSIFIED).

3. *Ibid.*, p. 21.

4. N. Friedman, *U.S. Naval Weapons*, U.S. Naval Institute Press, Annapolis, MD, 1983, p.134 (UNCLASSIFIED).

5. Klein, "Notes," p. 48.

6. R. F. DelSanto, Jr., "Early Developments in the Field of Sonar (1900-1939)," NUSL Technical Memorandum 904.3-1-64, Navy Underwater Sound Laboratory, New London, CT, 8 January 1964, p. 12 (UNCLASSIFIED).

7. Friedman, *U.S. Naval Weapons,* p. 257.

---

*When a reference is used in more than one chapter, it is repeated in full in the subsequent chapter(s). If the source is cited again *in the immediately succeeding note* within the same chapter, the designation *Ibid.* is used (see chapter 2, reference 3). However, if there are intervening sources before it is mentioned again, the name of the author and a short title are used (see chapter 2, reference 5).

Chapter 11 — Annotated Endnotes

CHAPTER 2: HISTORICAL BACKGROUND (Cont'd)
Pages 5–18

8. *Ibid.*, pp. 257-259.

9. M. Lasky, "Historical Review of Underwater Acoustic Technology: 1939-1945 with Emphasis on Undersea Warfare," *U.S. Navy Journal of Underwater Acoustics,* vol. 25, October 1975, pp. 890-892 (UNCLASSIFIED).

10. Friedman, *U.S. Naval Weapons,* p. 259.

11. T. G. Bell, "Sonar and Submarine Detection," NUSL Report 545, Navy Underwater Sound Laboratory, New London, CT, 1962 (UNCLASSIFIED).

12. "Scanning Sonar Systems," Summary Technical Report of Division 6, National Defense Research Committee, Washington, DC, 1946 (UNCLASSIFIED).

13. Friedman, *U.S. Naval Weapons,* p. 263.

14. *Ibid.*, p. 137.

15. S. E. Morison, *The Two-Ocean War*, Little, Brown and Company, Boston, 1963, reprinted by Galahad Books, New York, p. 561.

16. A description of the Walther power plant and the post-war British experimentation with it appears in R. Compton-Hall, *Submarine Versus Submarine*, Orion Books, New York, 1988, pp. 24-26. A discussion of the pros and cons of various air-independent propulsion (AIP) systems may be found in A. J. Donaldson, "Submarine Power Sources for the Mission," *Naval Engineers Journal*, May 1996, pp. 129-146 (UNCLASSIFIED).

17. N. Polmar "How Many Submarines?" *Proceedings of the Naval Institute*, February 1998, p. 87 (UNCLASSIFIED). Rear Admiral Momsen's estimate proved to be pessimistic. Polmar states that U.S. Naval Intelligence reported the Soviet submarine force to have peaked at some 450 active units about 1957. Their post-World War II construction of 727 submarines through 1991 was still formidable at an average submarine building rate of 16 per year, almost four times that of the United States.

18. Documents discovered after the collapse of East Germany indicate that the concern was well founded. A description of a major 1983 exercise (*Soyuz*-83) revealed Soviet plans to reach the French border 13 to 15 days after war broke out. A special road network had been built in East Germany to support a three-pronged attack, with vehicles and rolling stock suitable for Western roads and railroads stockpiled. See N. Friedman, *"The Fifty Year War: Conflict and Strategy in the Cold War,"* U.S. Naval Institute Press, Annapolis, MD, 2000, pp. 424-425 (UNCLASSIFIED). As recently as 1988, M. J. Gouge published a paper on the seriousness of the Soviet

Chapter 11 — Annotated Endnotes

CHAPTER 2: HISTORICAL BACKGROUND (Cont'd)
Pages 5–18

submarine threat for shipping lanes to Europe. See "Soviet Subs Vs. the Resupply of NATO," *Proceedings of the Naval Institute*, December 1988, pp. 109-114 (UNCLASSIFIED).

19. M. A. Palmer referred to the results of the Navy study in *Origins of the Maritime Strategy: The Development of American Naval Strategy, 1945-1955*, U.S. Naval Institute Press, Annapolis, MD, 1988, p. 69 (UNCLASSIFIED).

20. D. Middleton, *Submarine, the Ultimate Naval Weapon — Its Past, Present, and Future,* Playboy Press, Chicago, 1976, p. 130.

21. R. J. Urick, *Principles of Underwater Sound,* McGraw-Hill, Inc., Third Edition, New York, 1983, p. 148.

22. Morison, *Two-Ocean War*, p. 562.

23. J. Merrill, *Fort Trumbull and the Submarine*, Publishing Directions, Avon, CT, 2000.

24. Urick, *Principles.*

25. R. J. Urick, "The Utilization of Acoustic Paths in Long-Range Sonar Search," NRL Progress Report, Naval Research Laboratory, Washington, DC, November 1951, pp. 1-4 (UNCLASSIFIED). I am indebted to Dr. Burton G. Hurdle for locating the above report at NRL and for arranging to have it declassified.

26. F. E. Hale, "Long-Range Sound Propagation in the Deep Ocean," *Journal of the Acoustical Society of America*, vol. 33, no. 4, April 1961, pp. 456-464.

27. J. W. Horton, "Requirements of a 40-Kiloyard Echo-Ranging System," NUSL Report 221, Navy Underwater Sound Laboratory, New London, CT, 20 January 1954 (UNCLASSIFIED). This calculation technique is based on the simplifying assumption that the background consists only of noise.

28. H. W. Marsh, Jr., and M. Schulkin, "Report on the Status of Project AMOS (1 January 1953 to 31 December 1954)," NUSL Report 255, Navy Underwater Sound Laboratory, New London, CT, 21 March 1955 (UNCLASSIFIED).

29. H. L. Saxton, H. R. Baker, and N. Shear, "10 Kilocycle Long-Range Search Sonar," NRL Report 4515, Naval Research Laboratory, Washington, DC, 19 August 1955 (UNCLASSIFIED).

Chapter 11 — Annotated Endnotes

CHAPTER 3. LAUNCHING A LONG-RANGE ACTIVE SONAR PROGRAM: EARLY CONCEPT FORMULATION
Pages 19–60

1. J. W. Horton, "Requirements of a 40-Kiloyard Echo-Ranging System," NUSL Report 221, Navy Underwater Sound Laboratory, New London, CT, 20 January 1954 (UNCLASSIFIED).

2. M. Schulkin, "History, Development, and Present Status of Project AMOS (Acoustic, Meteorological, and Oceanographic Survey)," NUSL Report 132, Navy Underwater Sound Laboratory, New London, CT, 20 April 1951, pp. 7-13 (UNCLASSIFIED).

3. T. G. Bell, "Fundamental Design Considerations for a Reliable, Long-Range, Echo-Ranging Sonar," NUSL Research Report No. 301, Navy Underwater Sound Laboratory, New London, CT, 5 March 1956 (UNCLASSIFIED).

4. J. W. Horton, *Fundamentals of Sonar,* U.S. Naval Institute Press, Annapolis, MD, 1957, pp. 317-324 (UNCLASSIFIED).

5. "Scanning Sonar Systems," Summary Technical Report of Division 6, National Defense Research Committee, Washington, DC, 1946, pp. 45-46 (UNCLASSIFIED).

6. The expression was attributed to James Kyle, former head of the NUSC Submarine Sonar Department (see J. Merrill and L. D. Wyld's *Meeting the Submarine Challenge: A Short History of the Naval Underwater Systems Center,* United States Government Printing Office, 1997, p. 49 (UNCLASSIFIED)). Kyle had the expression mounted on a plaque in his office. In a conversation, he stated that the expression probably came from the radar field but could not remember its exact origin. In any event, Kyle popularized the expression at NUSC and helped drive home an important fundamental of the sonar design process.

7. Commander C. Graham, "What Every Subsystem Engineer Should Know About Ship Design — But Does Not Ask," *Naval Engineers Journal,* June 1978, pp. 83-91 (UNCLASSIFIED). See also C. Graham, "The Impact of Subsystems on Naval Ship Design," *Naval Engineers Journal,* December 1975, pp.15-25 (UNCLASSIFIED). Although Graham did not mention the ship class to which his calculation pertained, in the context of the articles, it appears to be a *Spruance* class with a 7100-ton displacement. Graham later (as a navy captain) would be the commanding officer of the David Taylor Research Center.

8. T. G. Bell, "Travel to CNO and NRL to Discuss Sonar Ship," NUSL Technical Memorandum 1230-0123-55, Navy Underwater Sound Laboratory, New London, CT, 8 August 1955 (UNCLASSIFIED).

Chapter 11 — Annotated Endnotes

CHAPTER 3. LAUNCHING A LONG-RANGE ACTIVE SONAR PROGRAM: EARLY CONCEPT FORMULATION (Cont'd)
Pages 19–60

9. Bell, "Fundamental Design Considerations."

10. W. A. Downes, "A Brief History of the AN/SQS-26 Project at NUSC New London Laboratory," NUSC Technical Memorandum SA2-0198-71, Naval Underwater Systems Center, New London, CT, 2 November 1971, p. 2 (UNCLASSIFIED).

11. J. L. Stewart, E. C. Westerfield, and M. K. Brandon, "Optimum Frequencies for Sonar Detection," *Journal of the Acoustical Society of America,* vol. 33, 1961, p. 1216.

12. J. L. Stewart, E. C. Westerfield, and M. K. Brandon, "Optimum Frequencies for Noise-Limited Active Sonar Detection," *Journal of the Acoustical Society of America*, vol. 70, 1981, p. 1336.

13. W. H. Thorp, "Deep Ocean Sound Attenuation in the Sub and Low Kilocyle-Per-Second Region," *Journal of the Acoustical Society of America*, vol. 38, 1965, pp. 648-654. See also W. H. Thorp, "Analytical Description of the Low-Frequency Attenuation Coefficient," *Journal of the Acoustical Society of America*, vol. 42, 1967, pp. 270-271.

14. R. H. Mellen, P. M. Scheifele, and D. G. Browning, "Global Model for Sound Absorption in Sea Water," *Scientific and Engineering Studies*, Naval Underwater Systems Center, New London, CT, 1987 (UNCLASSIFIED).

15. W. A. Downes, "Report of Travel to BuShips on 14 November 1955," NUSL Technical Memorandum 1230-0189-55, Navy Underwater Sound Laboratory, New London, CT, 28 November 1955 (UNCLASSIFIED).

16. F. S. White, "Specification Material for a Long Range, Active, Search Sonar," NUSL Technical Memorandum 1230-023-56, Navy Underwater Sound Laboratory, New London, CT, 31 January 1956 (UNCLASSIFIED).

17. T. G. Bell, "Report of BuShips Conference on Sonar Scout Ship," NUSL Technical Memorandum 1230-017-56, Navy Underwater Sound Laboratory, New London, CT, 26 January 1956 (UNCLASSIFIED).

18. T. G. Bell, "Comments on a Bottom Bounce Proposal," NUSL Technical Memorandum 1230-045-56, Navy Underwater Sound Laboratory, New London, CT, 24 February 1956 (UNCLASSIFIED).

19. M. Muir, Jr., *Black Shoes and Blue Water: Surface Warfare in the United States Navy, 1945-1975*, Naval Historical Center, Washington, DC, 1996, p. 79 (UNCLASSIFIED).

Chapter 11 — Annotated Endnotes

CHAPTER 3. LAUNCHING A LONG-RANGE ACTIVE SONAR PROGRAM: EARLY CONCEPT FORMULATION (Cont'd)
Pages 19–60

20. A preliminary study was drafted 16 April 1958 and later published as T. G. Bell's "A Theoretical Comparison of the Bottom-Bounce Capabilities of AN/SQS-23 Versus AN/SQS-26 Surface-Ship Sonars," NUSL Report 449, Navy Underwater Sound Laboratory, New London, CT, 15 October 1959 (UNCLASSIFIED).

21. T. E. Thuma, "Report on Factors Affecting the Reliability of Sonar Transducer Elements" (U), General Electric Company/Heavy Military Systems, Syracuse, NY, 29 April 1969 (CONFIDENTIAL).

22. S. A. Peterson, "Report of Travel to Washington D.C. on 7, 8, and 9 February 1956," NUSL Technical Memorandum 1200-014-56, Navy Underwater Sound Laboratory, New London, CT, 27 February 1956 (UNCLASSIFIED).

23. Bell, "Fundamental Design Considerations."

24. H. J. Morrison, "Bottom Bounce Sonar Cylindrical Transducer and Beam Patterns," NUSL Technical Memorandum 1230-0165-56, Navy Underwater Sound Laboratory, New London, CT, 14 August 1956 (UNCLASSIFIED). Also see J. F. Kelly, "Computed Beam Patterns for a Cylindrical Transducer for Bottom Bounce Sonar," NUSL Technical Memorandum 1110-024-56, Navy Underwater Sound Laboratory, New London, CT, 29 August 1956 (UNCLASSIFIED).

25. O. A. Hahs, "Report of Visit to BuShips on 10 & 11 July 1956," NUSL Technical Memorandum 1230-0145-56, Navy Underwater Sound Laboratory, New London, CT, 23 July 1956 (UNCLASSIFIED).

26. H. J. Wilms, "Details of Proposed Sea Tests Related to Long Range Bottom Bounce Echo Ranging Planned for March 1956," NUSL Technical Memorandum 1210-08-56, Navy Underwater Sound Laboratory, New London, CT, 13 January 1956 (UNCLASSIFIED).

27. H. J. Wilms, "Preliminary Results of DIC4A Sea Tests — Weeks of 19 & 26 March 56," NUSL Technical Memorandum 1210-040-56, Navy Underwater Sound Laboratory, New London, CT, 5 April 1956 (UNCLASSIFIED).

28. Hahs, "Visit to BuShips."

29. H. J. Wilms, "Preliminary Results of Scaled Down Bottom Bounce Submarine Echo Ranging Tests (Project BRASS)," NUSL Technical Memorandum 1210-088-56, Navy Underwater Sound Laboratory, New London, CT, 14 November 1956 (UNCLASSIFIED).

CHAPTER 3. LAUNCHING A LONG-RANGE ACTIVE SONAR PROGRAM: EARLY CONCEPT FORMULATION (Cont'd)
Pages 19–60

30. R. V. Lewis, "Proposed Experimental Equipment for Performing Bottom Bounce Sonar Investigations Aboard Submarines in the Deep Ocean," NUSL Technical Memorandum 1210-035-57, Navy Underwater Sound Laboratory, New London, CT, 1 March 1957 (UNCLASSIFIED).

31. R. V. Lewis, "BRASS II Status Report," NUSL Technical Memorandum 1210-042-58, Navy Underwater Sound Laboratory, New London, CT, 10 April 1958 (UNCLASSIFIED).

32. H. J. Wilms, "BRASS II — An Experimental Long Range Bottom-Reflected Active Sonar System," NUSL Technical Memorandum 1210-095-59, Navy Underwater Sound Laboratory, New London, CT, 1 December 1959 (UNCLASSIFIED). A somewhat expanded version of the same material by the same author appeared as NUSL Report 483, 24 August 1960 (UNCLASSIFIED).

33. H. W. Marsh, Jr., and M. Schulkin, "Report on the Status of Project AMOS (1 January 1953 to 31 December 1954)," NUSL Report 255, Navy Underwater Sound Laboratory, New London, CT, 21 March 1955, pp. 33-39 (UNCLASSIFIED).

34. T. G. Bell, "Bottom Loss Measurements and Range Predictions," NUSL Technical Memorandum 1230-041-1958, Navy Underwater Sound Laboratory, New London, CT, 1 April 1958 (UNCLASSIFIED).

35. T. G. Bell, "Discussion of Surface Vessel Sonar Problems at BuShips and CNO," NUSL Technical Memorandum 1230-011-57, Navy Underwater Sound Laboratory, New London, CT, 16 January 1957 (UNCLASSIFIED).

36. H. J. Morrison, "A Proposed Beam Forming and Switching Network for Bottom-Bounce Sonar," NUSL Technical Memorandum 1230-032-57, Navy Underwater Sound Laboratory, New London, CT, 19 February 1957 (UNCLASSIFIED).

37. The most articulate discussion on the problem with the CW waveform for a zero-Doppler target may be found in J. T. Kroenert, "Discussion of Detection Threshold with Reverberation-Limited Conditions," *Journal of the Acoustical Society of America,* vol. 71, no. 2, 1982, pp. 507-508.

38. M. Schulkin, F. S. White, Jr., and R. A. Spong, "QHBa Figure of Merit Tests," NUSL Report 187, Navy Underwater Sound Laboratory, New London, CT, 3 April 1953 (UNCLASSIFIED).

39. T. G. Bell, "Proposal for Increasing the Scan Rate of Long-Range Search Systems," NUSL Report 146, Navy Underwater Sound Laboratory, New

Chapter 11 — Annotated Endnotes

CHAPTER 3. LAUNCHING A LONG-RANGE ACTIVE SONAR PROGRAM: EARLY CONCEPT FORMULATION (Cont'd)
Pages 19–60

London, CT, 28 January 1952 (UNCLASSIFIED); also see by the same author "Optimum Use of a Cylindrical Multi-Element Transducer for Antisubmarine Search," NUSL Report 170, Navy Underwater Sound Laboratory, New London, CT, 23 September 1952 (UNCLASSIFIED).

40. T. G. Bell, "Directional Transmission as a Method of Increasing the Source Level of Cylindrical, Multi-Element Transducer for Cavitation Limited Performance," NUSL Report 159, Navy Underwater Sound Laboratory, New London, CT, 13 June 1952 (UNCLASSIFIED).

41. R. E. Baline, "Report of Travel to NEL, 10-12 June 1957," NUSL Technical Memorandum 1230-0152-57, Navy Underwater Sound Laboratory, New London, CT, 9 October 1957 (UNCLASSIFIED).

42. J. L. Stewart *et al.* (authors' names unavailable), "A Comparison of the Detection Capabilities of Single Frequency, Frequency Modulated, and Noise Sonars for LORAD," NEL Technical Memorandum 70, Naval Electronics Laboratory, San Diego, CA, 28 December 1954 (UNCLASSIFIED). The underlying theory was later published in the open literature by J. L. Stewart and E. C. Westerfield, "A Theory of Active Sonar Detection," *Proceedings of the IRE*, May 1959 (UNCLASSIFIED).

43. The proposal to use a combined FM-CW transmission turned out not to be completely original. In 1964, Alvin Novick of Yale University provided the first description of a similar FM-CW biosonar in the mustached bat. The bat uses the CW part of the signal for initial detection of flying insects, with its Doppler processing capability separating an insect echo from echoes in the surrounding vegetation. As the bat closes range, it relies on its FM waveform and associated processing for precision ranging, target imaging, and target motion analysis, with a processing technique equivalent to a pulse-compression, matched-filter system. This astonishing processing capability is accomplished in a brain about the size of a pea. The Novick reference and other information on bat sonar is discussed by Nobuo Sugo in "Biosonar and Neural Computation in Bats," *Scientific American*, June 1990, pp. 60-68.

44. Navy Underwater Sound Laboratory letter to Bureau of Ships, Serial 1230-02, 4 January 1957 (UNCLASSIFIED).

45. H. E. Nash, "Trip to BuShips and CNO for Consideration of the Bottom-Bounce/First Convergence Zone Sonar System Procurement," NUSL Technical Memorandum 1200-04-57, Navy Underwater Sound Laboratory, New London, CT, 17 January 1957 (UNCLASSIFIED).

46. Downes, "Brief History of AN/SQS-26 Project at NUSC."

Chapter 11 — Annotated Endnotes

CHAPTER 3. LAUNCHING A LONG-RANGE ACTIVE SONAR PROGRAM: EARLY CONCEPT FORMULATION (Cont'd)
Pages 19–60

47. R. E. Baline and W. A. Downes, "Conclusion on Development Philosophies for SQS-26 Sonar," NUSL Technical Memorandum 1230-086-58, Navy Underwater Sound Laboratory, New London, CT, 23 May 1958 (UNCLASSIFIED).

48. W. A. Downes, "Report of Visit to BuShips on 23 May 1958," NUSL Technical Memorandum 1230-077-58, Navy Underwater Sound Laboratory, New London, CT, 26 May 1958 (UNCLASSIFIED).

49. R. E. Baline, "Report of Travel to BuShips to Discuss GE's AN/SQS-26 Proposal," NUSL Technical Memorandum 1230-0111-58, Navy Underwater Sound Laboratory, New London, CT, 11 July 1958 (UNCLASSIFIED).

50. Downes, "Brief History of AN/SQS-26 Project at NUSC."

51. R. E. Baline, "Report of Meeting with EDO Corporation on 9 July 1958," NUSL Technical Memorandum 1230-0136-58, Navy Underwater Sound Laboratory, New London, CT, 13 August 1958 (UNCLASSIFIED).

52. Letter from H. Tucker, General Electric, to T. Bell, commenting on an early draft of this book, 14 November 1999 (UNCLASSIFIED).

53. V. C. Anderson, "DELTIC Correlator," Technical Memorandum 37, Harvard Acoustics Laboratory, Cambridge, MA, 5 January 1956 (UNCLASSIFIED).

54. R. E. Baline, "Report of Travel to Naval Electronics Laboratory," NUSL Technical Memorandum 1230-0148-58, Navy Underwater Sound Laboratory, New London, CT, 5 September 1958 (UNCLASSIFIED).

55. W. A. Downes, "Report of Travel to BuShips on AN/SQS-26," NUSL Technical Memorandum 1230-0109-59, Navy Underwater Sound Laboratory, New London, CT, 19 June 1959 (UNCLASSIFIED).

56. N. Friedman, *U.S. Naval Weapons*, U.S. Naval Institute Press, Annapolis, MD, 1983, p. 110 (UNCLASSIFIED).

57. W. A. Downes, "Notes on a Discussion with Elmer Landers on SQS-26 Installation Plans as of 31 March 1959," NUSL Technical Memorandum 1230-057-59, Navy Underwater Sound Laboratory, New London, CT, 1 April 1959 (UNCLASSIFIED).

58. N. Polmar, *The Ships and Aircraft of the U.S. Fleet,* U.S. Naval Institute Press, Fourteenth Edition, 1987, pp. 170-179 (UNCLASSIFIED).

Chapter 11 — Annotated Endnotes

CHAPTER 3. LAUNCHING A LONG-RANGE ACTIVE SONAR PROGRAM: EARLY CONCEPT FORMULATION (Cont'd)
   Pages 19–60

59. Downes, "Brief History of AN/SQS-26 Project at NUSC."

60. Muir, *Black Shoes and Blue Water,* p. 131.

61. B. S. Blanchard and W. J. Fabrycky, *Systems Engineering and Analysis*, Third Edition, Prentice Hall, Inc., Englewood Cliffs, NJ, 1998, p. 23.

62. J. B . Fisk, "The Bell Telephone Laboratories," *The Organization of Research Establishments*, J. Cockcroft, ed., Cambridge University Press, London, 1965, pp. 205-208. In addition to diplomatically interfacing with the research organization, NUSL found dealing with the *development* departments even more of a challenge because this function involved a critique what was being done in those departments. On the other hand, maintaining independence from the development work appeared essential to the systems engineering function.

63. T. G. Bell, "ASW Control of Large Ocean Areas with Mobile Forces," NUSL Report 474, Navy Underwater Sound Laboratory, New London, CT, 6 May 1960 (UNCLASSIFIED).

64. The concept bears some resemblance to one proposed in a different context by Churchill during World War I for the visual search of the German commerce raider *Emden*: "It is no use stirring about the oceans with two or three ships. When we have got Cruiser sweeps of 8 or 10 vessels ten or fifteen miles apart, there will be some good prospect of utilizing information as to the whereabouts of the *Emden* . . . ." From W. S. Churchill, *The World Crisis*, vol. 1, Barnes and Noble, New York, 1993, pp. 255-256.

65. R. J. Urick, *The Utilization of Acoustic Paths in Long-Range Sonar Search*, NRL Progress Report, Naval Research Laboratory, Washington, DC, November 1951 (UNCLASSIFIED).

CHAPTER 4. FULL-SCALE EXPERIMENTATION AND DEVELOPMENT
   Pages 61–90

1. D. W. Tufts and A. J. Cann, "On Albersheim's Detection Equation*,"* *IEEE Transactions on Aerospace and Electronic Systems (Letters),* vol. AES-19, no. 4, 1983, pp. 643-645.

2. E. H. Ahlstrom and R. C Counts,. "Eggs and Larvae of the Pacific Hake Merluccius Productus," *Fishery Bulletin 99,* vol. 56, U.S. Department of the Interior Fish and Wildlife Service, 1955, pp. 295-329.

CHAPTER 4. FULL-SCALE EXPERIMENTATION AND DEVELOPMENT (Cont'd)
Pages 61–90

3. W. A. Downes, "A Brief History of the AN/SQS-26 Project at NUSC New London Laboratory," NUSC Technical Memorandum SA2-0198-71, Naval Underwater Systems Center, New London, CT, 2 November 1971 (UNCLASSIFIED).

4. W. A. Downes, "Report of Meetings at BuShips on the SQS-26 Project on 2 April 1963," NUSL Technical Memorandum 930-86-63, Navy Underwater Sound Laboratory, New London, CT, 3 April 1963 (UNCLASSIFIED).

5. T. G. Bell, "Observations of AN/SQS-26 (XN-1) Performance on the USS *Willis A. Lee* (DL-4) from 16 to 23 August 1963," NUSL TM 905-083-63, Navy Underwater Sound Laboratory, New London, CT, 27 August 1963 (UNCLASSIFIED).

6. Downes, "Brief History of AN/SQS-26 Project at NUSC."

7. L. T. Einstein, "A Review of the U.S. Navy Underwater Sound Laboratory's Iso-Loss Contour Program," NUSL Report No. 708, Navy Underwater Sound Laboratory, New London, CT, 19 October 1965 (UNCLASSIFIED).

8. A. F. Wittenborn, "Analysis of Signal Processing and Related Topics Pertaining to the AN/SQS-26 Sonar Equipment," Tracor Summary Report III delivered under BuShips Contract Nobsr-93140, Tracor Inc., Austin, TX, 11 October 1965, p. 3 (UNCLASSIFIED).

9. D. E. Weston, "Correlation Loss in Echo Ranging," *Journal of the Acoustical Society of America*, vol. 37, no.1, January 1965.

10. J. L. Stewart and M. K. Brandon, "Random Medium? Correlation Loss?" Presented at the *NATO-Marina Italian Advanced Study Institute on Stochastic Problems in Underwater Sound Propagation*, Lercici, Italy, 18-23 September 1967 (UNCLASSIFIED).

11. W. A. Downes, "Conclusions Resulting from a 4 November 1969 Meeting on SQS-26 Display Problems," NUSL TM 2130-760-69, Navy Underwater Sound Laboratory, New London, CT, 1 November 1969 (UNCLASSIFIED).

12. M. Schulkin and H. W. Marsh, "Shallow Water Transmission," *Journal of the Acoustical Society of America,* vol. 34, no. 6, 1962, pp. 863-864.

Chapter 11 — Annotated Endnotes

CHAPTER 5. PROTOTYPE TESTING
Pages 91–98

1. R. J. Doubleday, "An Examination of Existing Communications Between the Fleet and Naval Laboratories" (U), NUSC Technical Memorandum RA42-C29-74, Naval Underwater Systems Center, New London, CT, 8 August 1974 (CONFIDENTIAL).

2. J. R. Hill, *Anti-Submarine Warfare*, Second Edition, U.S. Naval Institute Press, Annapolis, MD, 1989, p. 115 (UNCLASSIFIED).

CHAPTER 6. SUPPORTING RESEARCH AND DEVELOPMENT
Pages 99–110

1. J. M. Young, "Marking Density Studies for the AN/SQS-26 Sonar Equipment A-Scan Display," Tracor Document 66-316-U, Tracor Inc., Austin, TX, 20 May 1966 (UNCLASSIFIED).

2. J. M. Young, "The Effect of Signal Detectability of Using Four, Five, Six, or Seven Discrete Marking Intensity Levels," Tracor Document 67-649-U, Tracor Inc., Austin, TX, 16 August 1967 (UNCLASSIFIED).

3. G. T. Kemp, "Some Computed Effects of Fresh Water in the Sonar Dome of the AN/SQS-26," Tracor Document 66-351-U, Tracor Inc., Austin, TX, 11 June 1966 (UNCLASSIFIED).

4. R. P. Chapman and J. R. Marshall, "Low Frequency Reverberation from Deep Scattering Layers in the Western North Atlantic," Paper G-6, *22nd U.S. Navy Symposium on Underwater Acoustics*, October 1964 (UNCLASSIFIED).

5. R. P. Chapman and J. H. Harris, "Surface Backscattering Strengths Measured with Explosive Sound Sources," *Journal of the Acoustical Society of America,* vol. 34, 1962, p. 1592.

6. T. G. Bell, "Discussion with R. P. Chapman of NRE on Surface and Biological Backscattering," NUSL Technical Memorandum 905-058-64, Navy Underwater Sound Laboratory, New London, CT, 1964 (UNCLASSIFIED); also see B. F. Cole and T. G. Bell, "Further Comments on Biological Backscatter," NUSL Technical Memorandum 905-094-64, Navy Underwater Sound Laboratory, New London, CT, 2 November 1964 (UNCLASSIFIED).

7. R. P. Chapman, O. Z. Bluy, R. H. Adlington, and A. E. Robison, "Deep Scattering Layer Spectra in the Atlantic and Pacific Oceans and Adjacent Seas," *Journal of the Acoustical Society of America,* vol. 56, 1974.

CHAPTER 6. SUPPORTING RESEARCH AND DEVELOPMENT (Cont'd)
Pages 99–110

8. J. B. Hersey and R. H. Backus, "New Evidence That Migrating Gas Bubbles, Probably the Swimbladders of Fish, Are Largely Responsible for Scattering Layers on the Continental Rise South of New England," *Deep-Sea Research*, vol. 1, 1954, pp. 190-191.

9. R. H. Backus and J. B. Hersey, "The Geographical Variation of Midwater Sound Scattering" (U), Report 66-10, Woods Hole Oceanographic Institution, Woods Hole, MA, March 1966 (CONFIDENTIAL).

10. R. L. Boivin and B. Thorpe, "Results of Measurements on SQS-26 Display System," NUSL Technical Memorandum 934-442-64, Navy Underwater Sound Laboratory, New London, CT, 9 December 1964 (UNCLASSIFIED).

11. R. W. Young, "Example of Propagation of Underwater Sound by Bottom Reflection," *Journal of the Acoustical Society of America*, vol. 20, no. 4, July 1948, pp. 455-462.

12. T. G. Bell, "Predicting and Dealing with Energy Spreading Loss," presented at the Navy Underwater Sound Laboratory, New London, CT, and published in the *Proceedings of a Seminar on Active Sonar Signal Processing*, Section 3, Tracor Document T90-01-9532-U, Tracor Inc., Austin, TX, 12 December 1989 (UNCLASSIFIED).

13. R. H. Mellen and D. Browning, "Variability of Low-Frequency Sound Absorption in the Ocean: pH Dependence," *Journal of the Acoustical Society of America*, vol. 61, no. 3, March 1977, pp. 704-706.

14. T. G. Bell, "Range Prediction for the AN/SQS-26," NUSL Technical Memorandum 905-082-65, Navy Underwater Sound Laboratory, New London, CT, 21 December 1965 (UNCLASSIFIED).

15. E. M. Podeszwa, "Sound Speed Profiles for the North Pacific," NUSC Technical Document 5271, Naval Underwater Systems Center, New London, CT, 2 February 1976 (UNCLASSIFIED); also see, by the same author, "Sound Speed Profiles for the North Atlantic," NUSC Technical Document 5447, 20 October 1976 (UNCLASSIFIED); "Sound Speed Profiles for the Indian Ocean," NUSC Technical Document 5555, 11 December 1976 (UNCLASSIFIED); "Sound Speed Profiles for the Norwegian Sea," NUSC Technical Document 6035, 4 June 1979 (UNCLASSIFIED); and "Sound Speed Profiles for the Mediterranean Sea," NUSC Technical Document 6309, 15 August 1980 (UNCLASSIFIED).

Chapter 11 — Annotated Endnotes

CHAPTER 7. THE RUBBER DOME WINDOW
Pages 111–120

1. W. A. Downes, "Some Observations Regarding Dome Paint Problems with AN/SQS-23 and AN/SQS-26 Ships," NUSL Technical Memorandum 930-146-62, Navy Underwater Sound Laboratory, New London, CT, 2 July 1962 (UNCLASSIFIED).

2. J. O. Natwick, "Acoustic Measurements of Rubber Window Materials for Sonar Domes," NUSL Technical Memorandum 2133-628-69, Navy Underwater Sound Laboratory, New London, CT, 15 September 1969 (UNCLASSIFIED).

3. S. Anthopolos, Jr., "Results of Self-Noise Level Measurements Made on 31 July 1972 Aboard USS *Bradley* (DE-1041) with a B. F. Goodrich Bow-Mounted Rubber Sonar Dome Window," NUSC Technical Memorandum SA22-0131-72, Naval Underwater Systems Center, New London, CT, 15 August 1972 (UNCLASSIFIED).

4. W. S. Burdic, *Underwater Acoustic Systems Analysis*, First Edition, Prentice-Hall, Englewood Cliffs, NJ, 1984, p. 345. I am indebted to Walter Hay, formerly of NUSC, for calling this reference to my attention.

5. T. G. Bell, G. A. Brown, and E. M. Podeszwa, "An Analysis of the Expected Operational Effectiveness of Alternative Active Sonar Concepts for Surface Ships," NUSC Technical Report 4837, Naval Underwater Systems Center, New London, CT, 29 October 1974 (UNCLASSIFIED).

6. E. M. Podeszwa, *Convergence Zone Slide Rule (Revised)*, Naval Underwater Systems Center, New London, CT, June 1973 (UNCLASSIFIED).

CHAPTER 8. EQUIPMENT OPERATION AND TACTICAL EMPLOYMENT
Pages 121–128

1. J. R. Hill, *Anti-Submarine Warfare*, Second Edition, U.S. Naval Institute Press, Annapolis, MD, 1989, pp. 110-115 (UNCLASSIFIED).

2. R. W. Hunter, *Spy Hunter,* U.S. Naval Institute Press, Annapolis, MD, 1999, pp. 202-203 (UNCLASSIFIED). In the 1960's, NUSL frequently carried out testing on *Wilkinson* some 700 miles off the coast of Florida in an area off the shipping lanes. It was surprising one day to see a Soviet intelligence trawler suddenly appear on a closing course with *Wilkinson*. The trawler finally passed close astern to *Wilkinson*, presumably trying to determine if anything was being towed. No doubt the trawler was also manning listening hydrophones to determine *Wilkinson's* sonar transmitting

Chapter 11 — Annotated Endnotes

CHAPTER 8. EQUIPMENT OPERATION AND TACTICAL EMPLOYMENT (Cont'd)
Pages 121–128

frequencies and wave shapes (and was probably — as someone then remarked — sending pictures of *Wilkinson* to Moscow). At the time, it was not clear how the Soviets knew *Wilkinson* was there. The United States was unaware that the Soviets, with the help of cryptologic and keys provided by either Walker or other spies, were routinely reading "secure" Navy messages containing exercise plans and test locations. Before Walker began to work for the Soviets, Joseph Helmich, starting in 1963, sold details of the Navy code machines to the Soviets and provided key lists for 2 years after that. Some observers have speculated that there could have well been other unmasked spies operating during the 1960's and later. See J. Bamford, "The Walker Espionage Case," *Naval Institute Proceedings, Naval Review Issue*, 1986, pp. 110-119 (UNCLASSIFIED).

CHAPTER 9. FLEET PERFORMANCE
Pages 129–170

1. H. Adams, *The Education of Henry Adams*, Oxford World's Classics, Oxford University Press, Inc., New York, 1999, p. 253.

2. N. R. Augustine, *Augustine's Laws and Major System Development Programs*, American Institute of Aeronautics and Astronautics, Inc., New York, 1982, pp. 99-101.

3. N. Polmar, *Guide to the Soviet Navy*, Second Edition, U.S. Naval Institute Press, Annapolis, MD, 1986, p. 42 (UNCLASSIFIED).

4. H. J. Doebler, "Deep Sound Velocity Profiles in the Mediterranean Sea," NUSL Technical Memorandum 2040-72-68, Navy Underwater Sound Laboratory, New London, CT, 16 August 1968 (UNCLASSIFIED).

5. H. J. Doebler, "Hydrographic Data for the Mediterranean Sea," NUSL Publication 1021, Navy Underwater Sound Laboratory, New London, CT, 18 April 1969 (UNCLASSIFIED).

6. J. G. Keil and R. C. Chapman, "Improvements in Fleet Utilization of the AN/SQS-26 Sonar: Progress Report No. 4," NUSC Technical Report 4553, Naval Underwater Systems Center, New London, CT, 1 June 1973 (UNCLASSIFIED).

7. *Ibid.*

8. R. H. Mellen, "Global Model for Sound Absorption in Sea Water: Impact Study, Part 1," Final Report, OMNI Analysis, Waterford, CT, April 1988

Chapter 11 — Annotated Endnotes

CHAPTER 9. FLEET PERFORMANCE (Cont'd)
Pages 129–170

(UNCLASSIFIED); prepared for NUSC Code 61 under Contract N66604-87-D-0090.

9. T. G. Bell, "Seamounts as Submarine-Like Active Sonar Targets," Final Report, Vitro Corporation, New London, CT, 30 September 1988 (UNCLASSIFIED); prepared for R. L. Mason, NUSC Code 3315.

10. Soon promoted to Captain, Myers was to masterfully exploit the convergence zone capabilities of the SQS-26. In preparation for assignment in the Mediterranean, he spent an intense several days at the Navy Underwater Sound Laboratory in New London, CT, learning about the capabilities of the SQS-26. He documented his views on the tactical use of the convergence zone mode in "ASW in the Mediterranean: An Operator's Report to the Operators" (U), *COMASWFORLANT ASW Quarterly*, Spring 1970 (CONFIDENTIAL). Myers advanced to rear admiral in 1971 and retired in 1979. During his career, he was awarded the Defense Superior Service Medal, The Legion of Merit (with two Gold Stars in lieu of second and third awards), and the Navy Commendation Medal.

11. B. F. Cole, "AN/SQS-26 Shallow Water Echo-Ranging Results in the Mediterranean Sea," NUSC Technical Report 4401, Naval Underwater Systems Center, New London, CT, 12 October 1972 (UNCLASSIFIED). Because a "truncated" detection is one that occurs immediately at the start of a closing run, it does not provide an accurate test of the maximum possible detection range that would occur if the submarine had started at a greater initial range.

12. J. J. Hanrahan, E. M. Podeszwa, F. E. Rembetski, and T. O. Fries, "Bottom Bounce Echo-Ranging Trials of the AN/SQS-26 (AXR) in the Eastern Mediterranean," NUSC Technical Report 4555, Naval Underwater Systems Center, New London, CT, 26 June 1973 (UNCLASSIFIED).

13. Keil and Chapman, "Fleet Utilization of AN/SQS-26 Sonar: Progress Report 4."

14. *Ibid.*

15. J. G. Keil and R. C. Chapman, "Improvements in Fleet Utilization of the AN/SQS-26 Sonar: Progress Report No. 5," NUSC Technical Report 4669, Naval Underwater Systems Center, New London, CT, 28 December 1973 (UNCLASSIFIED).

16. *Ibid.*

17. J. G. Keil, M. S. Hoyt, and R. S. Janus, "Improvements in Fleet Utilization of the AN/SQS-26 Sonar: Progress Report No. 6," NUSC Technical

CHAPTER 9. FLEET PERFORMANCE (Cont'd)
Pages 129–170

Report 5235, Naval Underwater Systems Center, New London, CT, 7 January 1976 (UNCLASSIFIED).

18. T. G. Bell, "Performance Trends in Active Sonar," NUSC Technical Memorandum PA3-C86-73, Naval Underwater Systems Center, New London, CT, 7 December 1973 (UNCLASSIFIED).

19. R. A. Connole, *USS Connole (FF-1056) History, August 30 1969 to August 30 1992,* 30 August 1992 (UNCLASSIFIED). The author, Rickart A. Connole, was the son of Commander David R. Connole, after whom the *Connole* was named. Connole, a highly decorated commanding officer of the *Trigger* (SS-237), was lost in action with *Trigger* off the coast of Japan in March 1945.

20. Keil *et al*, "Fleet Utilization of AN/SQS-26 Sonar: Progress Report 6."

CHAPTER 10. CONCLUSIONS
Pages 171–184

1. W. A. Downes, "A Brief History of the AN/SQS-26 Project at NUSC New London Laboratory," NUSC Technical Memorandum SA2-0198-71, Naval Underwater Systems Center, New London, CT, 2 November 1971, p. 8 (UNCLASSIFIED).

2. O. R. Cote, "The Third Battle: Innovation in the U. S. Navy's Silent Cold War Struggle with Soviet Submarines," March 2000, p. 22, http:www.chinfo navy mil/navpalib/ships/submarines/sub100.html.

3. A. V. Kuteinikov, "Malachite Subs Post Proud Tradition," *Naval Institute Proceedings*, April 1998, p. 55 (UNCLASSIFIED).

4. R. J. Murray, "Russia's Threat Beneath the Surface," *Wall Street Journal*, 25 August 1995.

5. J. R. Hill, *Anti-Submarine Warfare*, Second Edition, U.S. Naval Institute Press, Annapolis, MD, 1989 (UNCLASSIFIED).

6. "The Future Mix of Subs and Strategy," *U.S. Naval Institute Professional Seminar Series*, U.S. Naval Institute, Annapolis, MD, 25 September 1986, p. 8 (UNCLASSIFIED).

7. W. D. O'Neil, "Winning the ASW Technology Race," *U.S. Naval Institute Proceedings*, October 1988, p. 87 (UNCLASSIFIED).

8. J. L. Frank III, "An *Arleigh Burke* with Helicopters," *U.S. Naval Institute Proceedings*, October 1992, p. 102 (UNCLASSIFIED).

CHAPTER 10. CONCLUSIONS (Cont'd)
Pages 171–184

9. N. Polmar, "A Continuing Interest . . . in Submarines," *U.S. Naval Institute Proceedings*, November 1994, p. 103 (UNCLASSIFIED).

10. *Technology for the United States Navy and Marine Corps, 2000-2035, Volume 7: Undersea Warfare*, Naval Studies Board, National Research Council, Washington, DC, National Academy Press, 1997, p. 30 (UNCLASSIFIED).

APPENDIX A

CHRONOLOGY OF EVENTS INFLUENCING THE DEVELOPMENT AND APPLICATION OF THE SQS-26 SONAR

Appendix A

Year	Closely Associated with SQS-26 Program	Independent of SQS-26 Program
1914		Chilowsky, a Russian, proposes underwater acoustic echo-ranging for finding submarines.
1915		In February, Langevin, a French physicist, begins to implement Chilowsky's concept in Paris.
1916		In March, Langevin completes development of experimental underwater sound equipment and achieves one-way transmission across the Seine.
1918		Langevin, in a controlled experiment off Toulon, receives echoes from a submarine.
1923		NRL is established in Washington, DC, and Langevin's work is continued there. The first Superintendent of NRL's Sound Division is H. Hayes.
1927		NRL completes and installs the first echo-ranging sonar (QA) on a destroyer. One-mile submarine detection range is achieved off Key West, FL. Eight QA sonars are eventually used on destroyers.
1945		HUSL, under the direction of F. V. Hunt, completes scanning sonar development. N. Friedman calls it "a connecting link to all post-war systems."
1948		NRL initiates field work on long-range surface duct and bottom-reflected paths. Bottom path loss is found to be insensitive to target depth and thermal conditions.

Note: In this appendix, CZ means convergence zone and BB represents bottom bounce.

Appendix A

Year	Closely Associated with SQS-26 Program	Independent of SQS-26 Program
1948 Cont'd		The QHB scanning sonar enters Fleet. Previous sonars covered only a narrow sector per ping. The QHB was the first production sonar to cover 360° per ping.
		The Soviets announce their intention to construct a fleet of 1200 submarines by 1965.
1949		NEL initiates field experiments on long-range CZ paths. CZ path loss is found to be little affected by target depth.
		NUSL initiates a propagation measurement program (AMOS) on surface duct and bottom-reflected paths in vital sea lanes of the North Atlantic.
1950		The SQS-10 scanning sonar enters the Fleet.
1951		Soviets commission first *Whiskey* submarine. By 1957, 236 *Whiskey* submarines will be in service, representing an enormous peacetime building rate of 36 per year.
		In November, NRL's R. Urick describes how long-range surface duct and bottom-reflected paths in the deep ocean might be tactically used for convoy protection.
1952		USS *Dealy* (DE-1006), 1817 tons, is laid down. This first post-war escort ship will carry the new SQS-4 scanning sonar. Only 13 *Dealy* class ships will enter service.

Appendix A

Year	Closely Associated with SQS-26 Program	Independent of SQS-26 Program
1954		The SQS-4 enters the Fleet.
		In January, NUSL's J. W. Horton describes the characteristics required for 40-kyd echo-ranging systems that could use either BB or surface duct paths.
1955	NUSL's W. Downes assigns T. Bell the task of determining what should be the next step in sonar development for surface ships.	In January, the first nuclear-propelled submarine, USS *Nautilus*, departs from New London, CT, for initial sea trials.
	During early summer, Bell completes conceptual design of a surface ship sonar for echo-ranging to 40 kyd via the BB path.	D. Middleton (a military correspondent for the *New York Times*) assesses *Nautilus* impact: "[T]he destroyer's old prey became the hunter, not only of other submarines but of the destroyers themselves."
	During July visit to OPNAV-312, Bell learns of "scout ship" proposal by CDR L. O'Brien for a specialized ASW escort ship capable of mounting a large sonar.	In March, NUSL's H. Marsh and M. Schulkin publish the first detailed bottom reflection propagation analysis versus frequency and angle from AMOS propagation loss data.
	In July, Bell visits NRL's H. Saxton, who is disappointed with NRL's BB echo-ranging.	In May, NEL demonstrates CZ echo-ranging with the LORAD system on submarine USS *Baya* (AGSS-318).
	In September, the NUSL BB sonar proposal is presented to NRL and BuShips personnel.	
	In September, BuShips requests that NUSL provide specifications for a new surface ship sonar by the end of December.	

A-5

Appendix A

Year	Closely Associated with SQS-26 Program	Independent of SQS-26 Program
1955 Cont'd	In November, during a USAG-sponsored symposium at University of Pennsylvania, NUSL's Bell delivers paper on BB sonar concept for an ASW escort ship.	
	In November, Downes and Bell attend meeting at BuShips regarding CDR O'Brien's scout ship.	
	In December, NUSL's H. Wilms outlines proposed submarine sea experiments supporting a long-range, BB echo-ranging program.	
1956	In January, Wilms plans tests with sub-mounted BRASS (Bottom-Reflected Active Sonar System) to support NUSL BB echo-ranging program.	ONI study announces that since World War II the USSR has built more cruisers, destroyers, and submarines than the rest of the world combined.
	In January, Bell attends meeting on the preliminary results of the scout ship study at BuShips, Code 420. CNO wants presentation of results in late January.	
	In January, NUSL's F. White drafts initial specifications for a sonar with a mechanically steerable rectangular array. Specifications are based on Bell's paper delivered at USAG-sponsored symposium in November 1955.	
	In January, Downes directs NUSL's H. Morrison to look at an electrically steered fixed cylindrical array in place of the mechanically steered rectangular array proposed in the Bell paper.	
	In February, Bell recommends against a BuShips proposal to use an SQS-23 type system with beam tilting for the scout ship BB sonar.	

Appendix A

Year	Closely Associated with SQS-26 Program	Independent of SQS-26 Program
1956 Cont'd	In February, NUSL (at a CNO Plans and Policy Group meeting) recommends that the NUSL sonar design be used for the scout ship sonar. Recommendation is favorably received.	
	In March, Bell publishes NUSL report documenting his USAG November 1955 symposium paper. BB ranges from 20 to 40 kyd are expected, along with CZ ranges at ~60 kyd.	
	In March, Wilms completes BRASS tests against surface craft echo-repeater target that show BB path is best for poor thermal conditions.	In June, NEL publishes LORAD summary report on experiments with CZ echo-ranging.
	In July, Morrison's cylindrical array with has been accepted by BuShips. A dome design has been provided by BuShips, Code 420.	
	In August, Morrison documents the beamforming network required to electrically tilt the beam.	
	Later in August, NUSL's J. Kelly computes the horizontal and vertical patterns for the phasing and shading specified by Morrison.	
	In September, Wilms successfully conducts first NUSL attempt at BB echo-ranging on a submarine target with BRASS in a "scaled test" at 100-800 fathoms.	
	In December, Bell finds support at CNO for interlaboratory committee. CNO is *leaning toward prototype ship versus experimental (scout) ship as easier to sell, foretelling scout ship demise.*	

Appendix A

Year	Closely Associated with SQS-26 Program	Independent of SQS-26 Program
1956 Cont'd	In December, the initial BB sonar specification is revised by NUSL to require an electrically steered, cylindrical transducer array and CZ capability.	
1957	In January, a specification for both BB *and CZ performance with a cylindrical array* is forwarded by NUSL to BuShips.	
	In January, NUSL's CAPT H. Ruble and H. Nash discuss (at BuShips) funding for developmental system and obtain agreement to form an inter-laboratory committee to meet 28-29 January.	
	In February, Morrison proposes beamforming and switching design for BB/CZ cylindrical array sonar.	In August, off Vladivostok, Soviet surface ships surround USS *Gudgeon*, which is sighted after broaching and is held down to exhaustion before finally surfacing.
	In September, BuShips issues specification for BB/CZ sonar to be designated the "SQS-26."	In October, USSR launches "Sputnik" satellite, astonishing the world with a demonstration of advanced Soviet technology.
1958	In April, *NUSL's R. Baline proposes the type of combined FM-CW transmission and processing on SQS-26 sonar.*	SQS-23 becomes operational with a nominal range in surface duct of about 12 kyd. Ultimately, the Navy procures 197 SQS-23's.
	In May, the policy is formulated to concentrate on BB for an SQS-26 EDO system with an early delivery. GE would deliver a more complete BB/CZ SQS-26 system later.	The first Soviet nuclear submarine is commissioned.
	In June, *contracts are placed for an SQS-26 (XN-1) with EDO and for an SQS-26 (XN-2) with GE.* Combined FM-CW waveforms are added shortly after signing.	

A-8

Appendix A

Year	Closely Associated with SQS-26 Program	Independent of SQS-26 Program
1958 Cont'd	In June, the first meeting is held at NEL with BuShips, GE, EDO, and NUSL representatives to discuss the design and performance of the NEL LORAD sonar.	
1959	Long Range Objectives Group (OP-93) asks BuShips to investigate building inexpensive escorts with SQS-26 sonar and companion weapon.	In May, USS *Grenadier* (SS-525) chases Soviet *Zulu* for 9 hours off Iceland, forcing *Zulu* to surface. As the first ASW success versus Soviets, *Grenadier* wins case of Jack Daniels from CINCLANT.
	In August, the CNO Ship's Characteristics Board decides that *all new Fleet escorts should mount the SQS-26 sonar*. It is too large to backfit on existing ships.	In August, Bell leaves Surface Ship Sonar Branch for staff group in the Sonar Development Division. Staff group provides systems engineering function for fixed, submarine, and surface sonars.
	In October, Bell formally publishes detailed analysis on why Sangamo's proposed use of the SQS-23 as a BB sonar makes no sense.	
	In December, Wilms *describes first deep-water BB echo-ranging tests with BRASS II*. Volume reverberation but no surface reverberation is seen.	
1960	Production contract is let to GE for two SQS-26 sonars. These are the first production systems and will be installed on USS *Bronstein* (DE-1037) and USS *McCloy* (DE-1038).	The 2560-ton *Bronstein* and *McCloy* (DEs, later designated FFs) are authorized. *These are the first ships to carry production versions of the SQS-26 sonar.*
	In May, Bell describes a method of employing a large-area SQS-26 sweep to detect and neutralize ballistic-missile-firing submarines.	In November, USS *George Washington* (SSBN-598), the first Polaris submarine, puts to sea in Atlantic for first patrol, with a full load of 16 Polaris missiles.
1961	EDO's experimental SQS-26 (XN-1) is installed on USS *Willis A. Lee* (DL-4), a cruiser-size ship, for technical evaluation under project T/S 25.	USS *Thresher* (SSN-593), considered to be "the first quiet SSN," is completed.

Appendix A

Year	Closely Associated with SQS-26 Program	Independent of SQS-26 Program
1961 Cont'd	Production contract is let to GE for 12 AX sonars (prior to obtaining echo-ranging results from the SQS-26 (XN) experimental systems).	*Garcia* class (DEs, later FFs) of 10 ships and *Brooke* class (DEGs, later FFGs) of 6 ships are authorized. These 3400-ton ships received either the SQS-26 (AX) or (BX).
		In September, F. Hale of NEL publishes first paper in the open literature on CZ propagation.
1962	Production contract is let to EDO for 18 BX sonars.	The FY62-FY67 budgets fund 60 surface warships designed principally for ASW, a fivefold increase in number for this type of ship over the previous decade.
	GE's SQS-26 (XN-2) is installed on USS *Wilkinson* (DL-5) for technical evaluation under project T/S 26.	Nine *Belknap* DLG/CGs at 7940 tons and one *Truxton* CGN at 9127 tons are authorized. These cruisers would receive either the SQS-26 (AX) or (BX).
	In June, the first SQS-26 submarine echoes are received via the BB path with SQS-26 (XN-1) on *Lee*. *Reverberation is discovered to dominate background.*	
	In November, the first SQS-26 submarine echoes are received via a CZ path with the SQS-26 (XN-2) on *Wilkinson*.	
1963	In February, B. F. Goodrich proposes the inflatable rubber dome as an alternative to the existing steel bulbous bow dome.	USN Warrant Officer J. G. Helmich, a communications specialist, sells details of the Navy's KL-7 crypto machine to the Soviet Union. He provides keys and other information for next 2 years.
	In March, Development Assist Project DS-281 is initiated to improve SQS-26 design. SOFIX office in BuShips would manage program. This is the first formal recognition of the need for more development.	H. Nash becomes the Technical Director of NUSL.

Appendix A

Year	Closely Associated with SQS-26 Program	Independent of SQS-26 Program
1963 Cont'd	In March, volume reverberation from the deep scattering layer off California is identified as a serious limitation to BB and CZ performance with the SQS-26 (XN-2).	
	In March, reverberation data from the SQS-26 (XN-1) indicate that that three 10° transmissions on adjacent bearings raise the received reverberation.	
	In April, Downes expresses concern with foreseeable problems in maintaining and operating the SQS-26 in the Fleet.	
	In May, NUSL's J. Snow recommends that ping-to-ping history storage replace existing "single-look" displays.	
	In June, the first SQS-26 storage display (for four beams) is tested on the SQS-26 (XN-1) aboard *Lee*.	
	In August, SQS-26 BB tracking firsts on XN-1 were reported as follows: (1) BB reception opening out to 50 kyd and (2) a BB closing run from 45 kyd at 10° to 10 kyd at 42°.	
	In September, the decision is made that all equipments after the AX will use a 40° stepped transmission beam for 120° coverage.	
	In October, the first analysis of BB surface reverberation on XN-2 shows agreement with predictions from the Chapman-Harris study. First nonbeam aspect BB tracking is achieved.	

Appendix A

Year	Closely Associated with SQS-26 Program	Independent of SQS-26 Program
1963 Cont'd	In December, first CW pulse tracking reception is demonstrated via BB path for high Doppler runs on XN-2 with good results.	
	In December, Development Assist Project DS-281 ends.	
1964	In January, an OPEVAL (OS-55) begins on *Wilkinson's* SQS-26 (XN-2).	Forty-six 3,877-ton *Knox* class ships (DEs, later FFs) are authorized. Plans call for installation of the SQS-26 (CX) on these ships.
	In May, the OPEVAL is terminated with the conclusion that the BB and CZ modes are unsatisfactory.	
	In May, CNO sets up a Navy-wide committee to review the soundness of the basic SQS-26 design as a result of the OPEVAL failure. The decision on acquisition of SQS-26 sonar for the *Knox* class is held up.	
	CNO committee membership includes T. Bell (NUSL), W. Carlson (TRW), CDR B. Becken (BuShips, 372), D. Andrews (NEL), CDR A. Glennon (OPTEVFOR), and CAPT W. Dobie (CNO).	
	In June, as a result of the SQS-26 CNO review, increased NUSL funding is provided for (1) design review, (2) XN-1 and XN-2 tests, (3) Fleet checkouts, and (4) operating guidance development.	
	In June, a second Development Assist project (DS-331) is established to investigate and correct SQS-26 problems revealed in testing to date.	

Appendix A

Year	Closely Associated with SQS-26 Program	Independent of SQS-26 Program
1964 Cont'd	In July, OPNAV review results in Navy proceeding with FY 64/65 procurements of 27 SQS-26 (CX) sonars for *Knox* class (DE-1052's).	
	In July, a brassboard with a three 40° stepped transmission beam scheme is installed on *Wilkinson* to provide 120° coverage.	
	In November, GE makes farfield receiving phase measurements on the array, indicating no inter-element coupling effects. Phases are as calculated.	In November, studies by R. P. Chapman of NRE, Halifax, give new perspective on relative seriousness of biological and surface scattering in the North Atlantic.
	In December, R. Boivin's 10- and 20-kyd display resolution measurements at NUSL indicate that the scan converter system limits resolution to about 65 ms.	
1965	In January, first XN-2 echo-ranging is conducted in the three 40° stepped transmission beam mode. Volume reverberation in Gulf of Mexico is found to degrade nighttime performance.	In the mid-1960's, the USSR becomes aware of the "need for increased acoustic secretiveness." *Measures for Soviet submarine noise reduction begin.*
	In January, Tracor's J. Collins makes recommendation on how to process explosive bottom loss data for use in SQS-26 performance predictions.	
	In March, contracts are awarded by the Naval Oceanographic Office to Texas Instruments and Alpine for a Marine Geophysical Survey (MGS), mainly of bottom loss in operationally important locations.	
	In March, Atlantic CZ propagation is found to have attenuation significantly greater than that in the Pacific. This mystery is explained by NUSC's R. Mellen and D. Browning in March 1976.	

A-13

Appendix A

Year	Closely Associated with SQS-26 Program	Independent of SQS-26 Program
1965 Cont'd	In May, first XN-2 shallow-water testing is conducted with an SQS-26. Time spreads, bottom loss, reverberation, and detection range are compared with models.	
	In May, Project DS-331 ends after nine sea test periods. Necessary changes are to be applied to the SQS-26, XN-2, AX, and CX designs.	
	May marks the end of major development work on the SQS-26 sonar equipment begun 7 years earlier with the awarding of contracts to EDO and GE in June 1958.	
	In May, GE's J. Costas reports 100- to 200-ms time spreading from analysis of tapes, which seemed to make high resolution undesirable. This is the genesis of proposed FSK waveform.	In September, NUSL's R. C. Chapman documents first SQS-26 free-play exercise results with *McCloy* in the surface duct mode. Detections are made from 10 to 22 kyd, with three of four ASW attacks being successful.
1966	In February, OPTEVFOR reports on evaluation of the surface duct mode of SQS-26 on *Bronstein*, concluding performance is at least equal to SQS-23.	
	In February, XN-2 biological reverberation change with time of day in the Gulf of Mexico is found to be the same as that near Bermuda. Bottom-limited focus zone is confirmed at low angles.	
	In February, NUSL's H. Fridge writes first "training requirements" technical memorandum for SQS-26.	In March, the Navy "bureaus" are abolished and replaced by naval system commands. BuShips becomes the Naval Sea Systems Command (NAVSEA).
	In April, Bell writes first operator manual for the SQS-26.	In March, Woods Hole Oceanographic Institution publishes the first study on the expected geographic variation of biological scattering.

Appendix A

Year	Closely Associated with SQS-26 Program	Independent of SQS-26 Program
1966 Cont'd	In April, the Goodrich experimental dome with a rubber window is installed on *Lee*.	
	In June, Bell reports the first detection testing for an SQS-26 (BX) production model on USS *Wainwright* (DLG-28).	In July, USS *Brooke* (DEG-1) with SQS-26 (AX) reports first free-play CZ detection by Fleet in exercise off San Diego.
	In August, noise measurements on *Lee* with the rubber dome window indicate an average improvement over a steel dome of about 5 dB.	
	In September, NUSL's R. Chapman conducts the first discussion between NUSL personnel and sonar school staff on the problem of teaching SQS-26 operation.	
	In December, analysis of CZ propagation loss in the North Atlantic test area indicates that the 1965 attenuation value of NUSL's W. Thorp is preferred over the AMOS value in use up to that time as the standard.	
1967	In April, a final report is issued on Tracor's study of the effect of the rubber dome on beam patterns and radiation impedance.	*Victor I* enters service in the Soviet Navy. It is the first in a series of quieter Soviet submarine designs.
	In August, NUSL's W. Hay reports on first sea tests (conducted under technical evaluation project T/S 51) for XN-2 major retrofit (MRF), which will be the prototype for production SQS-26's.	USN Warrant Officer John Walker, communications watch officer at COMSUBLANT, starts spying for the Soviet Union.
	In August, Tracor's F. Lagrone reports on a theoretical study relating to the apparent lack of any serious effect of array element interactions on transmitting beam patterns.	*John Walker reveals to USSR the success of U.S. Navy passive listening against Soviet submarines. This information is believed to be a contributing motivation to the Soviet submarine silencing effort.*

A-15

Appendix A

Year	Closely Associated with SQS-26 Program	Independent of SQS-26 Program
1967 Cont'd	In November, NUSL's H. Fridge and P. Cable report on an experimental study regarding the effect of the number of display echo-return histories on the detectability of an echo.	In July, NUSL's W. Thorp publishes an equation for North Atlantic attenuation based on his 1965 data and NUSL's D. Browning's fit showing a second relaxation frequency.
	In November, the scale model studies of NUSL's J. Hanrahan and General Dynamics' D. Nelson on a *Skipjack* class indicate the effect of target time spread on peak echo level as a function of resolution and aspect.	
1968	A production contract is let to GE for 27 more CX sonars.	
	In January, Tracor's J. Young issues a study on the relationship between marking density and detectability for seven levels and six echo cycles (0.5 to 0.8 was found to be a good marking density range).	
	In January, NUSL's C. Walker reports on a validation of MGS bottom loss province charts by comparing MGS predictions to SQS-26 sonar measurements.	
	In June, CAPT F. Kelly from the Sixth Fleet visits NUSL to discuss ASW assistance in the Mediterranean, given the poor near-surface thermal conditions that exist there.	
	In August, the seventh (and final) SQS-26 XN-2 (MRF) sea test is conducted, completing 6 years of XN-2 tests on Wilkinson.	
	In September, Bell, in response to a Sixth Fleet request, discusses Mediterranean CZ potential at ASWFORSIXTHFLT. NUSC assistance to the Mediterranean Fleet begins.	

Appendix A

Year	Closely Associated with SQS-26 Program	Independent of SQS-26 Program
1968 Cont'd	In October, NUSL's B. Cole conducts the first CZ echo-ranging demonstration in the Mediterranean under controlled conditions with an SQS-26 (BX). A total of 30 SQS-26's are now in the Fleet.	
	In October, Tracor's K. Hamilton reports on a study of SQS-26 transmitting beam patterns. It is concluded that the dreaded element interaction has no effect on the patterns.	
	In November, NUSL's J. Walsh reports that storage display tests indicate little difference in alerted versus unalerted operator performance.	In November, first wide-area search is performed with the SQS-26 in the Tyrrhenian Sea. Assistance is provided by Cole on USS *Koelsch* (DE-1049) and Bell on USS *Voge* (DE-1047).
1969	In January, NUSL's Cole and J. Hanrahan report on first analysis of bottom-scattering strengths in SQS-26 test areas of the Atlantic. Results are consistent with those of NEL's K. Mackenzie in the Pacific.	First ship of the *Knox* (DE-1052) class is commissioned. A total of 46 of this class would be commissioned between 1969 and 1974.
	In February, NUSL's W. Hay and T. Bell present (to COMOPTEVFOR) SQS-26 results on T/S 51 testing in the North Atlantic for the MRF version of the XN-2.	
	In March, Tracor's S. Fowler and J. Bednar conduct an analysis of MGS province 1 through 4 data, giving the difference between peak and total energy to be 8.5 plus 5 times the log sin of the grazing angle.	In March, NUSL's Bell and Chapman plan tactics and ride *Bronstein* for the first free-play carrier screening exercise by an SQS-26 ship. In 10 CZ opportunities with the AXR, 4 detections are made.
	In April, a GE study is completed of an SQS-26 mod that will include two new frequencies to alleviate mutual interference (MI).	In July, Cole (with the SQS-26) reports first use of the CZ mode in the Mediterranean by *McCloy* (with AXR) to track a Soviet submarine for a total continuous tracking time of 14 hours.

Appendix A

Year	Closely Associated with SQS-26 Program	Independent of SQS-26 Program
1969 Cont'd	In April, T. Thuma of GE writes a summary paper on nine sources of failures in SQS-26 transducer elements.	In July, *McCloy* in a free-play search of a 1° square in the Mediterranean detects an exercise submarine, first vectoring a VP and later an SQS-23 to attack range.
	In September, after NUSL's Chapman observes unexpected SQS-26 reverberation on Knox, he proposes a shipboard computer to predict range from in situ reverberation and bathythermographs — the genesis of SIMAS.	
	In September, NUSL's J. Natwick makes transmission loss measurements on 1.25-inch rubber panels of various constructions.	
	In November, NUSL's W. Downes summarizes the difficulty with the scan converter display storage tubes, with their lengthy 250-page alignment procedure.	
1970	In February, CNO (RADM L. O'Brien) calls a general training conference on the SQS-26, discussing sonar technical allocations, sonar courses, operating doctrine, etc. CDR R. Lage chairs.	
	In February, NUSC's J. Hanrahan and E. Podeszwa publish a discussion of the relationship between peak and total energy bottom loss in MGS provinces 1 through 4.	
	In May, Cole and Hanrahan present a military oceanography conference (MILOC) paper on the use of MGS for BB predictions. They also show biological reverberation versus time of day to be similar in the Mediterranean, Gulf of Mexico, and North Atlantic.	

Appendix A

Year	Closely Associated with SQS-26 Program	Independent of SQS-26 Program
1970 Cont'd	In August, NUSC's R. Chapman and J. Keil report on CZ testing at Guantanamo, comparing surface ship and submarine target strengths to justify the use of surface ships for CZ training.	
	In August, NUSC's Hanrahan conducts first controlled BB testing in Mediterranean with USS *Glover* (AGDE-1) and AXR. In 10 closing runs, he observes 9 detections at ranges out to 25 kyd.	In August, the Destroyer Development Group conducts free-play BB tests with *Glover* in the Mediterranean. In 10 closing runs, 6 detections are made.
	In September, NUSC's T. Chao provides a specification for providing higher opening-target Doppler coverage for the situation of a "stern chase" at high speed (noted by NUSC's Bell as a problem at sea in March 1969).	
	In November, NUSC's Chapman proposes a standardized multimode training plan for use aboard ships where a target is available.	
	In December, NUSC's Cole *et al.* compare bottom-scattering strength measurements among Mackenzie (NEL), Mediterranean MGS, and Mediterranean SQS-26 results.	
1971	Navy formally issues first SQS-26 Watch Supervisor's Manual (written by Chapman and Keil) to Fleet.	Surface ships begin to experiment with towed passive arrays for the protection of task forces.
	In May, Bell reports observations on the first "operational appraisal" cruise of the SQS-26 (CX) on USS *W. S. Sims* (DE-1059).	The 30-ship *Spruance* class (DD) is authorized.
	In August, Cole *et al.* report results for a program of CZ performance modeling tests in the Mediterranean (the JOAST program), which Cole planned and directed.	In April, USS *Horne* (DLG-30) with SQS-26 (BX), in a free-play HOLDEX exercise, makes nine separate CZ submarine detections, holding up to 2 hours, and vectors VP and surface units to datum.

Appendix A

Year	Closely Associated with SQS-26 Program	Independent of SQS-26 Program
1971 Cont'd	In August, NUSC's Chapman and Keil issue the first of a series of reports analyzing Fleet performance of the SQS-26 sonar.	
	In August-September, NUSC's Cole, in cooperation with La Spezia, conducts first SQS-26 shallow-water tests in the Mediterranean on the Tunisian Shelf with *Glover*. Median detection ranges of 15 kyd are observed.	In September, COMDESDEVGRU uses *Glover* in a free-play exercise against a submarine attempting penetration of a CZ coverage zone. Detection is made on 11 of 13 attempts. Average hold time is 48 min.
	In September, NUSC's Hanrahan, at the request of the Sixth Fleet, conducts BB echo-ranging tests with a submarine along a 1300-mile track from Malta to Egypt to validate NUSC bottom loss chart predictions.	In October, the Sixth Fleet conducts the first two-ship, free-play CZ sweep against an exercise submarine. Detection is achieved on each of two sweeps. One ship holds contact while vectoring VP or second ship to attack.
	In November, NUSC's Downes issues a brief history of the SQS-26 project.	In November, Downes retires from NUSC.
1972	In August, NUSC's S. Anthopolos reports first *Garcia* class rubber dome measurements, showing a major reduction of 15 dB over the standard steel dome noise.	*Victor II* enters service in the Soviet Navy with "a further decreased acoustic field."
	In September, OPTEVFOR issues the final report on an operational appraisal of the SQS-26 (CX) on *Sims* in the North Atlantic.	In April, *Sims* demonstrates SQS-26 (CX) capability to track a Soviet *Foxtrot* with an active sonar over a period of 8 days, using both surface duct and CZ paths.
	In October, Downes *et al.* issue a summary report on 17 years of SQS-26 development.	
	In November, NUSC's G. Brown issues a summary report on testing of the first computer-assisted shipboard performance prediction and mode selection system (SIMAS).	

Appendix A

Year	Closely Associated with SQS-26 Program	Independent of SQS-26 Program
1973	In March, NUSC's W. Roderick and R. Dullea issue a definitive report with both experimentation and theory on the use of explosive biological scattering data to predict CZ performance.	E. Yeager (Case Western Reserve University) identifies boric acid as being responsible for the second relaxation attenuation relation postulated by NUSL's D. Browning and W. Thorp in July 1967 (see July 1967 entry).
	In March, NUSC's Hay and Bell direct USS *Knox* (DE-1052)/USS *Kirk* (DE-1087) Pacific comparison tests to measure the CZ detection advantage from noise reduction due to the rubber dome. A major advantage is demonstrated.	In May, USS *Stein* (DE-1065) with SQS-26 (CX) demonstrates coordination with LAMPS and SQS-23 ship in CZ detection and attacks on exercise submarine in transit from Midway to Guam.
	In December, Bell compares the Fleet's surface duct performance of the steel dome SQS-26 to older sonars. Source is NUSC's D. Williams' data from ASW escort exercises in the North Atlantic.	In August, SHAREM XVI (MD) tests U.S. nuclear submarine's ability to penetrate three-ship CZ barrier. Depth is marginal for CZ. Rubber dome window ship detects 41% of attempts; steel domes, 15%.
1974		The last SQS-26 ship of DE-1052 class is commissioned, USS *Moinester* (DE-1097).
		Passive sonar in the U.S. Navy begins to become a dominant ASW detection technique — just as the last SQS-26 (CX) enters the Fleet.
1975		The destroyer escort (DE) name is changed to frigate (FF).
		Nash retires and N. Pryor becomes the new technical director at NUSC. Surface ship sonar is deemphasized as NUSC emphasis on the submarine mission is increased.
	In October, Bell and Hay direct LAMPS III development tests in the Atlantic to demonstrate rubber dome window SQS-26 ability to detect in CZ and vector a helo to carry out an attack.	

Appendix A

Year	Closely Associated with SQS-26 Program	Independent of SQS-26 Program
1976	From April-October, NUSC supports ASW Squadron operations in the Mediterranean.	In March, NUSC's Mellen and Browning publish a paper on Pacific versus Atlantic attenuation due to boric acid and pH differences. The CZ attenuation variation with location mystery is finally solved.
		During the summer, USS *Connole* (FF-1056), under CDR R. Pittenger, holds unalerted CZ active detection in CV screen on Soviet *Echo* II class submarine for 13 hours.
1977		In March, NUSC's Mellen and Browning publish a paper on the variability of the attenuation coefficient with worldwide location due to the variation in chemical characteristics of the sea water.
1978		*Victor III* enters service in the Soviet Navy with a noise level said to be "dozens of times less" than that of the first *Victor* submarine. Its noise level is considered comparable to USS *Permit*.
		First production SQR-18 passive array towed from variable depth sonar fish is installed on USS *Joseph Hewes* (FF-1078).
		Twenty-seven ships of the CG-47 class are authorized.
1982		First surface ship SQR-19 towed array is installed on USS *Moosbrugger* (DD-980).
1984		First *Akula* goes into service with further improvements in silencing.

Appendix A

Year	Closely Associated with SQS-26 Program	Independent of SQS-26 Program
1985		The Walker spy ring is discovered after some 18 years of operation inside the U.S. Navy.
		The *Arleigh Burke* class (DDG-51) is authorized.
1986		As reported at the Naval Institute Seminar in New London, CT, *there is recognition that the era of passive sonar as the primary ASW detection method is rapidly coming to an end.*
1988		In an October *Naval Institute Proceedings*, W. O'Neil states that the latest Russian nuclear submarines are detectable passively within only 5% of the range of earlier designs.
1989		As described by J. R. Hill in *Anti-Submarine Warfare* (second edition), *the 15-year dominance of passive sonar (since 1974) ends as the primary ASW detection method of the U.S. Navy*.
		In September, a *Naval Institute Proceedings* paper from the UK discusses the impact of quieter Soviet submarines on deployment and the decline in the U.S. Navy active capability because of years of reliance on passive sonar.
1991		Improved *Akula* enters service in the Soviet Navy. U.S. experts believe *Akula* to be quieter than the improved *Los Angeles* class.
		Gorbachev declares the Soviet Union to be extinct. The Cold War is over.

Appendix A

Year	Closely Associated with SQS-26 Program	Independent of SQS-26 Program
1992		In February, *the CNO Destroyer Variant Study deletes "towed array" from the FY94 DD-79 Flight IIA specification* (*Flight IIA is an upgrade to the DDG-51 design*).
1994		In a November *Naval Institute Proceedings* article, N. Polmar credits the Russians with the following statement: "The Americans lose the *Oscar II* immediately after the submarine puts out into the ocean!"
		In FY94, Flight IIA of the DDG-51 is authorized. The DDG-79, the first of 24, will join the Fleet in 2000.
1997		A National Research Council document on technology for the U.S. Navy says that ". . . passive detection ranges for these low-speed modern submarines have shrunk from hundreds of kilometers to only a few kilometers."
1998		In a *Naval Institute Proceedings* letter issued in December, it was said that ". . . in the future, ASW must recognize that passive acoustic data will not be available. Only active sensors will detect the submarine."

APPENDIX B

LIST OF PERSONNEL HAVING AN IMPACT ON THE SQS-26 PROGRAM

The following listing includes most of those individuals whose names appear in this memoir.

Appendix B

Alpine Geophysical Associates
Officer, C.

Bureau of Personnel (Navy)
Halley, J.

Bureau of Ships (later Naval Sea Systems Command)
Becken, Commander B.
Hanley, W.
Hudson, Commander W.
Kalina, Commander J.
Landers, E.
Mandel, P.
Priest, R.
Tiedeman, P.
Treitel, L.

Case Western Reserve University (Department of Chemistry)
Yeager, E.

Chief of Naval Material
Karaberis, Admiral C.

David Taylor Model Basin
Curtis, W.
Graham, Commander C.

Department of Defense Consultant
Friedman, N.

General Dynamics
Nelson, D.

General Electric
Costas, J.
Korolenko, K.
Sweetman, R.
Thuma, T.
Tucker, H.
Waful, L.

Harvard Underwater Sound Laboratory
Hunt, F. V.

Marine Physical Laboratory, Scripps
Anderson, V.

Massachusetts Institute of Technology
Cote, O.

Naval Electronics Laboratory
Anderson, E.
Andrews, D.
Barham, E.
Batzler, W.
Curl, G.
Hale, F.
Hamilton, E.
Mackenzie, K.
Pederson, M.
Stewart, J.
Westerfield, E.
Whitney, J.
Wilson, D.
Young, R.

Naval Oceanographic Office
Fry, Commander J.
Geddes, W.

Naval Research Establishment (Nova Scotia)
Chapman, R. P.

Naval Research Laboratory
Hayes, H.
Klein, E.
Saxton, H.
Urick, R.

Naval Ship Systems Command
Peale, Captain W. (SOFIX office)

Navy Underwater Sound Laboratory (Naval Underwater Systems Center, 1970)
Anthopolos, S.
Baline, R.
Bell, T.
Boivin, R.
Brown, G.
Browning, D.
Cable, P.
Chao, T.
Chapman, R. C.

Appendix B

NUSL/NUSC (Cont'd)
Clearwaters, W.
Cole, B.
DelSanto, R.
Doebler, H. J.
Downes, W.
Dullea, R.
Einstein, L.
Fisch, N.
Fridge, H.
Hanrahan, J.
Hay, W.
Holland, J.
Horton, J. W.
Keil, J.
Kelly, J.
Leibiger, G.
Lewis, R.
McFarland, Captain M.
Marsh, H.
Mellen, R.
Morrison, H.
Nash, H.
Natwick, J.
Pastore, M.
Peterson, S.
Podeszwa, E.
Roderick, W.
Ruble, Captain H.
Schulkin, M.
Sherman, C.
Silverio, A.
Snow, J.
Thorp, B.
Thorp, W.
Wainright, W.
Walker, C.
Walsh, J.
Wardell, W.
Whitaker, W.
White, F.
Williams, D.
Wilms, H.

Office of the Chief of Naval Operations
Dobie, Captain, W.
Glancey, Captain T.
Martell, Vice Admiral C.
Merrill, Captain S.

Metzel, Rear Admiral J.
O'Brien, Rear Admiral L.
Ricketts, Admiral C.
Rozier, Captain
Thach, Vice Admiral J.
Warder, Rear Admiral F.
Watkins, Admiral J.

Operational Test & Evaluation Force
Duggan, Lieutenant Commander R.
Glennon, Commander A.

Raytheon
Batchelder, L.

Texas Instruments
Horton, C. (consultant)

Tracor Inc.
Bednar, J.
Collins, J.
Fowler, S.
Hamilton, K.
Kemp, G.
Lagrone, F.
Wittenborn, A.
Young, J.

TRW
Carlson, W.

United States Navy
Kelly, Captain F.
Myers, Rear Admiral W.
Outlaw, Vice Admiral E.
Peale, Captain W.
Pittenger, Rear Admiral R.
Weschler, Rear Admiral T.
Zenni, Captain M.

Weapons System Evaluation Group (Staff of Secretary of Defense)
Bottoms, A.

Woods Hole Oceanographic Institution
Backus, R.
Ewing, M.
Hersey, J.

INDEX

Admiralty Underwater Weapons Establishment (AUWE), 109
Alpine Geophysical Associates, 103-105
AMOS (acoustic, meteorological, and oceanographic survey)
    Horton, use by, 16, 19
    NUSL initiation, 15
    versus BRASS results, 38
Anderson, Ernest, 29
Anthopolos, Savas, Jr., 112
Attenuation coefficient
    Atlantic versus Pacific, 107
    high values in Mediterranean, 137-139
Backus, Richard, 102
Baline, Russell
    FM/CW proposal, 46
    on core planning group, 30
    sea test responsibility, 79
Barham, Eric, 29, 66
Batzler, William, 29
Becken, Bradford
    as CDR, USN – key effectiveness study, 76
    as LCDR, USN – head of BuShips Sonar Code 848A, 49
    as LT, USN – in BuShips Sonar Code 848A, 48
Bell, Thaddeus
    active sonar use, decline of, 127
    and the ASW Squadron, 165
    Brooke (DEG-1) visit, 131-132
    carrier screening exercise, 143-144
    convergence zone path, inclusion of, 35
    CX tests, observation of, 96
    display fault discovery, 74
    echo excess study, 20-22
    Marine Geophysical Survey (MGS)
        concerns about, 103
        in performance prediction, 107-108
        steering committee, 104
    memoir, limitations on scope of, 172
    multiship sweep study, 55-57
    on core planning group, 30
    operational sweep (first), 139-142
    photograph of author, 142
    rubber window
        echo-ranging tests, 114-115
        impact study, 116-119
    sea mounts as false targets, 141-142
    sonar proposal, documentation of, 34

Index

Bell, Thaddeus (*continued*)
    SQS-26 guidelines responsibility, 80
    SQS-26 review committee, 76-78
    stepped-beam design (initial), 43-45
    symposium presentation, 27-29
    systems engineering, 52-55, 176
    training lectures, initial, 123
    XN-2 MRF tests, presentation on, 92
B. F. Goodrich Company, 111, 114
Biological reverberation
    Backus and Hersey, seminal work of, 102
    LORAD reverberation, seasonal influences on, 50
    on SQS-26 (XN-2) in Gulf of Mexico, 84
    worldwide studies by R. P. Chapman, 101
Boivin, Robert, 102-103
Bottom bounce path
    Bell study, 20-22
    BRASS experiments, 36-41
    description of, 2
    first free-play tests, 147
    Horton seminal study, 19
    normal incidence proposal, 41-42
    NRL experiments, 18
    Urick, early studies by, 14
    XN-1 and XN-2 early tests, 64-66
BRASS
    advantages and limitations, 62-64
    experimentation, 36-39
Brown, George, 108
Browning, David, 107
Bureau of Personnel (BuPers), 122
Bureau of Ships (BuShips)
    cylindrical array acceptance, 34-35
    initial contact with, 27
    Scout Ship meetings, 29-32
    SQS-26 procurement contracts, 48-51
Cable, Peter, 109
Canada, laboratory support by, 173, 175
Carlson, William, 76
Chapman, Richard
    carrier screening exercise, 143
    first free-play observations, 130
    Fleet ASW School, initial lectures at, 91
    operating doctrine program, directed by, 125
    shipboard prediction computer proposal, 108

Chapman, Robert
    biological and surface reverberation experiments, 101
    on MGS steering committee, 104
Chief of Naval Operations (CNO)
    Plans and Policy Group, 33-34
    scout ship, 25
    SQS-26 program support, 174
Chilowsky, Constantine, 5
Clearwaters, Walter, 36
Coded pulse waveforms
    a controversial topic, 63
    initial experimentation, 50
Cole, Bernard
    first Soviet submarine detection by SQS-26, 145
    first SQS-26 operational use in Mediterranean, 139
    first SQS-26 testing in Mediterranean, 137
    organizes JOAST, 110
    shallow-water testing in Mediterranean, 148
Collins, Joseph, 104
Colossus program, 89
COMASWFORSIXTHFLT, 132, 137
Convergence zone path
    carrier screening, use in, 143
    characteristics in Mediterranean, 135-136
    Connole tracking *Echo*-2, 166
    contacts reported from 1970 to 1974, 168
    decline in use after 1972, 154
    description of, 1-2
    exploitation by two SQS-26 ships, 150
    exploitation in Mediterranean, 132
    first detection of Soviet submarine with, 145
    first echo ranging by NEL, 17-18
    first echo ranging on operational ship, 65
    first free-play use, 131
    first helicopter vectoring with, 144
    first random detection of U.S. submarine, 151
    free-play exercises in Mediterranean, use in, 150
    HOLDEX 2-71 (Pacific), use in, 147-148
    LAMPS, use with in Pacific, 154-155
    Mediterranean attenuation impact on SQS-23, 139
    performance limitations (initial), 35-36
    rubber window
        capability beyond 70 kiloyards, 165
        capability with, 168
        impact of, 36
        superiority with, 156

Index

Convergence zone path (*continued*)
    Soviet submarine surveillance, use in, 152
    SQS-23 (Mediterranean), use with, 134
    SQS-26 design, exploitation of, 35
    U.S. submarine encounters, use in, 153
    WHOI, first studies at, 2
Cote, Owen, 181
Curl, Gilbert, 28
David Taylor Model Basin
    dome design, 24
    scout ship design, 31
DelSanto, Ralph, Jr., 6
Design review, 69
Destroyer Development Group
    first free-play bottom bounce tests, 147
    free-play convergence zone test, 150
Destroyer sonars
    QA, 6
    QB, 6-7
    QC, 6-7, 23
    QGB, 7
    QHB
        history, 7-9
        photograph of display, 8
    SQS-4, 9, 12, 32, 50
    SQS-10, 9
    SQS-23, 32, 51, 68, 74, 134
    SQS-26
        contract for experimental systems, 49
        first two production contracts, 49, 51, 68
        initial fleet performance, 130
        operational application, initial study of, 55-57
    SQS-26 (AX), 51, 68, 131, 151
    SQS-26 (AXR), 92, 143, 145-150, 152, 156, 166
    SQS-26 (BX), 51, 68, 91, 137, 139-141, 147-148, 166
    SQS-26 (CX), 51, 78, 96-98, 150-156, 165-166
    SQS-26 (XN-1)
        early testing on *Lee*, 64-65
        EDO contract, 49
        first storage displays, 71
        further testing on *Willis A. Lee*, 72
        T/S-25 technical evaluation on *Willis A. Lee*, 61
    SQS-26 (XN-2)
        display concerns, 50
        D/S-331 established, 80
        early testing on *Wilkinson*, 65-67

SQS-26 (XN-2) (*continued*)
    further testing on *Wilkinson*, 72-73
    GE contract, 49
    MRF version tests, 91-94
    operational evaluation, 73-75
    paper recorder display, 71
    refurbishing of, 70
    T/S-26 technical evaluation on *Wilkinson*, 61

Development assist projects
    D/S-281, 69, 90
    D/S-331, 80, 89-90

Display problems, 88

Doebler, Harold J.
    Naples assignment, 141
    SQS-23 convergence zone testing, 137

Downes, William A.
    bottom bounce sonar paper proposal, 27
    convergence zone capability, 35
    cylindrical array proposed, 34-35
    decision-making process, 30-31
    expanded SQS-26 program, description of, 78-80
    leadership ability, 180
    opening Doppler problem recognized, 144
    paper tape recorder, 50
    production model decisions, 51
    readiness for operational evaluation, 73
    shipboard training concerns, 70
    sonar displays, concern with, 88
    sonar study, initiation of, 19
    SQS-26 program accomplishments, 172
    steel dome problem analysis, 111

Echo excess
    advantage of, 21
    definition of, 20

EDO Corporation, 49, 180

Einstein, L. "Ted," 40

Ewing, Maurice, 2

Fault recognition, 83-84

Figure of merit, 15-16

Fisch, Norbert, 114

Fridge, Herbert
    display testing, 109
    training requirements study, 121

Geddes, Wilburt, 103

General Electric Company, 49, 50, 180

Graham, Clark (CDR, USN), 24

Index

Greenhalgh, Ken, 70
Hale, Frank, 29, 50
Hamilton, Edwin, 29
Hanley, William, 31, 42, 48
Hanrahan, John,
    first bottom bounce tests in Mediterranean, 146
    first reverberation analysis, 71
    Mediterranean bottom provinces, validation of, 149
    MGS, guidance on, 104
    sweep study, contribution to, 56
Harvard Underwater Sound Laboratory (HUSL)
    hull penetration constraint (21 inch), 23
    Hunt, F. V., director of, 5
    post-war merger with NUSL, 14
    scanning sonar, development of, 7
    scanning sonar, problems with, 43-44
Hay, Walter, 79, 81, 92
Hayes, Harvey
    active sonar, pioneer of, 16
    NRL sound division, head of, 6
    photograph of, 17
Hersey, J. Brackett, 102
Horton, J. Warren
    calculation methodology for performance, 21
    first bottom bounce sonar study, 16, 19
    optimum frequency concept, 22
    photograph of, 17
    reverberation, concern with, 20
    sweep concept, endorsement of, 56
Hunt, Frederick V.
    HUSL, director of, 5
    photograph of, 17
Kalina, John, (CDR, USN), 31
Keil, Juergen
    Mediterranean convergence zone anomaly, 137-138
    operating doctrine program, 124-125
Kelly, Fred (CAPT, USN), 133-134
Kemp, G. T., 99
Klein, Elias, 5
Korolenko, Kyrill, 70
Landers, Elmer, 31, 49, 69, 179
Langevin, Paul, 5
Leibiger, Gustave, 55
Lewis, Russell, 36-38
LORAD, 28, 46
McFarland, Milton (CAPT, USN), 114, 165

Index

Mackenzie, Kenneth, 29, 66
Marine Geophysical Survey (MGS), 103-107
Marsh, H. Wysor, 17
Martell, Charles (VADM, USN), 77, 78
Mellen, Robert, 29, 107, 138-139
Metzel, Jeffery (RADM, USN), 114
Myers, William (RADM, USN), 145
Morrison, Harold
    cylindrical array beamforming, 43
    cylindrical array specifications, 34-35
    display problems, 74
    sea testing, 79
    with core group, 30
Nash, Harold
    ASW Squadron, genesis of, 165
    attenuation experiments, initiation of, 29
    on "21-inch" mentality, 23
    NEL controversy, provoking of, 28
    normal-incidence bottom loss survey concept, 41
    NUSL Technical Director, 54
    rubber dome window test support, 114
    Sonar Systems Development Department, head of, 52
    steering committee proposal, 48
Natwick, Julius, 112
Naval Electronics Laboratory (NEL)
    convergence zone echo-ranging sonar, 17-18
    convergence zone studies, 15
    LORAD results, presentation of, 28
Naval Oceanographic Office, 103-107, 173, 175
Naval Research Laboratory (NRL), 6, 14, 25, 27, 48
Naval Sea Systems Command (NAVSEA), 174, 179
Naval Underwater Sound Laboratory (NUSL), 3, 14
Naval Warfare Publications, 125
O'Brien, Leslie (CDR, USN)
    scout ship, 29-30
    sonar ship proposal, 25-26
O'Brien, Leslie (RADM, USN), SQS-26 training conference, 124
Operational requirement problem, 31-32
Operational Test and Evaluation Force (OPTEVFOR)
    rubber window tests, observations of, 115
    SQS-26 (AX), operational appraisal of, 131
    SQS-26 (CX), operational appraisal of, 3, 96
    SQS-26 (XN-2), initial evaluation of, 73-75
    SQS-26 (XN-2) MRF, receives presentation on, 92
Peale, William (CAPT, USN)
    SOFIX office, head of, 69

Index

    training lectures, initiation of, 91
Pederson, Melvin, 29, 50
Peterson, Stanley, 30, 33, 52
Pittenger, Richard (CDR, USN), 166
Podeszwa, Eugene
    bottom loss domains, 106
    convergence zone slide rule, 119
    Mediterranean bottom loss chart, 147
    sound speed atlases, 108
Reliability
    management concerns, 68
    MRF and CX statistics, 93
    SOFIX program, 68-70
    XN-1 and XN-2 issues, 65, 67
Ricketts, Claude (ADM, USN), 77-78
Rubber dome window
    convergence zone performance, 168
    development, initiation of, 70
    importance of, 3
    in ASW Squadron, 165-166
    in LAMPS III tests, 165
    in SHAREM XVI (MD), 156
    initial testing, 111-119
    single most noteworthy gain, 171
SACLANT Undersea Research Center, 148, 173, 175
Sangamo Electric Company
    sonars built, 7-9
    SQS-23 versus SQS-26 controversy, 32-33
Saxton, Harold, 27
Schulkin, Morris
    adjustment of his bottom loss curves, 40-41
    AMOS analysis, 16-17
    detection probability methodology, 20-21
Scout ship
    BuShips design studies, 29-31, 34, 36
    concept demise, 42
    O'Brien's concept, 25-26
Search coverage evolution, 43-45
SEA teams, 94-96
Silverio, Albert, 130
Snow, John, 30, 71
Stewart, James, 28-29, 47, 50, 63, 88
Surface duct path
    described, 2
    in IEP exercises, 157-163
    seasonal variation in Mediterranean, 134-135

Index

Sweetman, Richard, 102
Systems engineering, 52-55
Technical evaluation projects
    T/S 25, 61, 89
    T/S 26, 61
    T/S 51, 92
Texas Instruments, 103-105
Thorp, Barry, 102
Thorp, William A., 29
Thuma, Theodore E., 33
Tiedeman, Paul, 179
Tracor, Inc.
    contributions, 180
    display history, effect of, 109
    recordings, analysis of, 82
    summary, publication of, 87
    supporting research and development, 69, 99-101, 104
Training
    early concerns of Downes, 70
    training conference, 124
    training lectures, initiation by Peale, 91
    training requirements study, 121
Urick, Robert
    below-layer shadow zone (figure), 11-12
    bottom bounce path, foreseen tactical use of, 14-15
    on MGS steering committee, 104
Wainwright, Walter, 79
Wardle, William, 29
Weschler, Thomas (RADM, USN), 36
Whitaker, Walter, 24
White, Frank S., 30, 79, 115
Williams, David, 157-163
Wilms, Hugo, 37-39
Wittenborn, Augustus, 87, 99
Woods Hole Oceanographic Institution (WHOI), 2, 102
Young, James, 100
Young, Robert, 104
Zenni, Marty (CAPT, USN), 139

PROBING THE OCEAN FOR SUBMARINES

A History of the AN/SQS-26 Long-Range, Echo-Ranging Sonar

Probing the Ocean for Submarines documents the key contributions made by numerous personnel and organizations to the AN/SQS-26 sonar development program. The nature of the technical problems encountered and the solutions found to address them are discussed, as well as the influence of international events on the objectives and support of the program. While the scientific challenges and international conditions have changed considerably over the years, the broad perspective offered in this book should be particularly helpful to those scientists and managers currently involved in naval research and development efforts.

About the Author

Thaddeus G. Bell, a 1945 graduate of Yale University with a B.S. degree in physics, spent 38 years in New London, Connecticut, at the Navy Underwater Sound Laboratory – in 1970 renamed the Naval Underwater Systems Center (NUSC). During that time, he was recognized as an international expert for submarine and surface ship sonar design, sea testing, performance analysis, shipboard performance predictions, and operating guidelines. Bell was especially well known to operators on submarines and destroyers for his sonar performance prediction publications. His most prominent accomplishment, however, was the conceptual design of the AN/SQS-26, which was the standard antisubmarine warfare sonar on destroyers and cruisers for over a decade.

After retiring from NUSC in 1985, Bell continued to work in sonar design and performance analysis for a number of corporations in the New England area.

Bell is a Fellow of the Acoustical Society of America and the recipient of many honors, including the Solberg Award from the American Society of Naval Engineers, the Navy Superior Civilian Service Award, and the David Bushnell Award from the American Defense Preparedness Association.

www.ingramcontent.com/pod-product-compliance
Lightning Source LLC
Chambersburg PA
CBHW082114230426
4367ICB00015B/2696